Crystallography
Made
Crystal Clear

Second Edition

Crystallography Made Crystal Clear: A Guide for Users of Macromolecular Models

Second Edition

Gale Rhodes
Department of Chemistry
University of Southern Maine
Portland, Maine
CMCC Home Page: www.usm.maine.edu/~rhodes/CMCC

ACADEMIC PRESS

San Diego   San Francisco   New York   Boston   London   Sydney   Tokyo

2006416414

This book is printed on acid-free paper. ∞

Academic Press
A Harcourt Science and Technology Company
525 B Street, Suite 1900, San Diego, California 92101-4495, USA
http://www.apnet.com

Academic Press
24-28 Oval Road, London NW1 7DX, UK
http://www.hbuk.co.uk/ap/

Library of Congress Catalog Card Number: 99-63088

International Standard Book Number: 0-12-587072-8

PRINTED IN THE UNITED STATES OF AMERICA
00 01 02 03 04 MM 9 8 7 6 5 4 3 2 1

Like everything, for Pam.

Contents

5. From Diffraction Data to Electron Density   85

6. Obtaining Phases   101

7. Obtaining and Judging the Molecular Model   133

Preface to the Second Edition

The first edition of this book was hardly off the press before I was kicking myself for missing some good bets on how to make the book more helpful to more people. I am thankful that heartening acceptance and wide use of the first edition gave me another crack at it, even before much of the material started to show its age. In this new edition, I have updated the first eight chapters in a few spots and cleaned up a few mistakes, but otherwise those chapters, the soul of this book's argument, are little changed. I have expanded and modernized the last chapter, on viewing and studying models with computers, bringing it up to date (but only fleetingly, I am sure) with the cyber-world to which most users of macromolecular models now turn to pursue their interests and with today's desktop computers—sleek, friendly, cheap, and eminently worthy successors to the five-figure workstations of the eighties.

My main goal, as outlined in the Preface to the First Edition, which appears herein, is the same as before: to help you see the logical thread that connects those mysterious diffraction patterns to the lovely molecular models you can display and play with on your personal computer. An equally important aim is to inform you that not all crystallographic models are perfect and that cartoon models do not exhaust the usefulness of crystallographic analysis. Often there is both less and more than meets the eye in a crystallographic model.

So what is new here? Two chapters are entirely new. The first one is "Other Diffraction Methods." In this chapter (the one I should have thought of the first time), I use your new-found understanding of X-ray crystallography to build an overview of other techniques in which diffraction gives structural clues. These methods include scattering of light, X rays, and neutrons by powders and solutions; diffraction by fibers; crystallography using neutrons and electrons; and time-resolved crystallography using many X-ray wavelengths at the same time. These methods sound forbidding, but their underlying principles are precisely the same as those that make the foundation of single-crystal X-ray crystallography.

The need for the second new chapter, "Other Types of Models," was much less obvious in 1992, when crystallography still produced most of the new macromolecular models. This chapter acknowledges the proliferation of such models from methods other than diffraction, particularly NMR spectroscopy and homology modeling. Databases of homology models now dwarf the Protein Data Bank, where all publicly available crystallographic and NMR models are housed. Nuclear magnetic resonance has been applied to larger molecules each year, with further expansion just a matter of time. Users must judge the quality of *all* macromolecular models, and that task is very different for different kinds of models. By analogies with similar aids for crystallographic models, I provide guidance in quality control, with the hope of making you a prudent user of models from all sources.

Neither of the new chapters contains full or rigorous treatments of these "other" methods. My aim is simply to give you a useful feeling for these methods, for the relationship between data and structures, and for the pitfalls inherent in taking any model too literally.

By the way, some crystallographers and NMR spectroscopists have argued for using the term *structure* to refer to the results of experimental methods, such as X-ray crystallography and NMR, and the term *model* for theoretical models such as homology models. To me, molecular structure is a book forever closed to our direct view, and thus never completely knowable. Consequently, I am much more comfortable with the term *model* for *all* of the results of attempts to know molecular structure. I sometimes refer loosely to a model as a *structure* and to the process of constructing and refining models as *structure determination,* but in the end, no matter what the method, we are trying to construct models that agree with, and explain, what we know from experiments that are quite different from actually looking at structure. So in my view, models, experimental or theoretical (an imprecise distinction itself), represent the best we can do in our diverse efforts to know molecular structure.

Many thanks to Nicolas Guex for giving to me and to the world a glorious free tool for studying proteins—SwissPdbViewer—along with plenty of support and encouragement for bringing macromolecular modeling to my undergraduate biochemistry students; for his efforts to educate me about homology modeling; for thoughtfully reviewing the sections on homology modeling; and for the occasional box of liqueur-loaded Swiss chocolates (whoa!). Thanks to Kevin Cowtan, who allowed me to adapt some of the clever ideas from his *Book of Fourier* to my own uses and who patiently computed image after image as I slowly iterated toward the final product. Thanks to Angela Gronenborn, Duncan McRee, and John Ricci for thorough, thoughtful, and helpful reviews of the manuscript. Thanks to Jonathan Cooper and Martha Teeter, who found and reported subtle and interesting errors lurking within figures in the first edition. Thanks to all those who provided figures—you are acknowledged alongside the fruits of your labors. Thanks to Emelyn Eldredge at Academic Press for inducing me to tiptoe once more through the minefields of Microsoft Word to update this little volume, and to Joanna Dinsmore for a smooth trip through production. Last and most, thanks to Pam for generous support, unflagging encouragement, and amused tolerance for over a third of a century. Time certainly does fly when we're having fun.

Gale Rhodes
Portland, Maine
March 1999

Preface to the First Edition

Most texts that treat biochemistry or proteins contain a brief section or chapter on protein crystallography. Even the best of such sections are usually mystifying— far too abbreviated to give any real understanding. In a few pages, the writer can accomplish little more than telling you to have faith in the method. At the other extreme are many useful treatises for the would-be, novice, or experienced crystallographer. Such accounts contain all the theoretical and experimental details that practitioners must master, and for this reason, they are quite intimidating to the noncrystallographer. This book lies in the vast and heretofore empty region between brief textbook sections on crystallography and complete treatments of the method aimed at the professional crystallographer. I hope there is just enough here to help the noncrystallographer understand where crystallographic models come from, how to judge their quality, and how to glean additional information that is not depicted in the model but *is* available from the crystallographic study that produced the model.

This book should be useful to protein researchers in all areas; to students of biochemistry in general and of macromolecules in particular; to teachers as an auxiliary text for courses in biochemistry, biophysical methods, and macromolecules; and to anyone who wants an intellectually satisfying understanding of how crystallographers obtain models of protein structure. This understanding is essential for intelligent use of crystallographic models, whether that use is studying molecular action and interaction, trying to unlock the secrets of protein folding, exploring the possibilities of engineering new protein functions, or interpreting the results of chemical, kinetic, thermodynamic, or spectroscopic experiments on proteins. Indeed, if you use protein models without knowing how they were obtained, you may be treading on hazardous ground. For instance, you may fail to use available information that would give you greater insight into the molecule and its action. Or worse, you may devise and publish a detailed molecular explanation based on a structural feature that is quite uncertain. Fuller understanding of the strengths and limitations of crystallographic models will enable you to use them wisely and effectively.

If you are part of my intended audience, I do not believe you need to know, or are likely to care about, all the gory details of crystallographic methods and all the esoterica of crystallographic theory. I present just enough about methods to give you a feeling for the experiments that produce crystallographic data. I present somewhat more theory, because it underpins an understanding of the nature of a crystallographic model. I want to help you follow a logical thread that begins with diffraction data and ends with a colorful picture of a protein model on the screen of a graphics computer. The novice crystallographer, or the student pondering a career in crystallography, may find this book a good place to start, a means of seeing if the subject remains interesting under closer scrutiny. But these readers will need to consult more extensive works for fine details of theory and method. I hope that reading this book makes those texts more accessible. I assume that you are familiar with protein structure, at least at the level presented in an introductory biochemistry text.

I wish I could teach you about crystallography without using mathematics, simply because so many readers are apt to throw in the towel upon turning the page and finding themselves confronted with equations. Alas (or hurrah, depending on your mathematical bent), the real beauty of crystallography lies in the mathematical and geometric relationships between diffraction data and molecular images. I attempt to resolve this dilemma by presenting no more math than is essential and taking the time *to explain in words what the equations imply.* Where possible, I emphasize geometric explanations over equations.

If you turn casually to the middle of this book, you will see some forbidding mathematical formulas. Let me assure you that I move to those bushy statement step by step from nearby clearings, making minimum assumptions about your facility and experience with math. For example, when I introduce periodic functions, I tell you how the simplest of such functions (sines and cosines) "work," and then I move slowly from that clear trailhead into the thicker forest of complicated wave equations that describe X rays and the molecules that diffract them. When I first use complex numbers, I define them and illustrate their simplest uses and representations, sort of like breaking out camping gear in the dry safety of a garage. Then I move out into real weather and set up a working camp, showing how the geometry of complex numbers reveals essential information otherwise hidden in the data. My goal is to help you see the relationships implied by the mathematics, not to make you a calculating athlete. My ultimate aim is to prove to you that the structure of molecules really does lie lurking in the crystallographic data—that, in fact, the information in the diffraction pattern implies a unique structure. I hope thereby to remove the mystery about how structures are coaxed from data.

If, in spite of these efforts, you find yourself flagging in the most technical chapters (4 and 7), please do not quit. I believe you can follow the arguments of these chapters, and thus be ready for the take-home lessons of Chapters 8 and 11, even if the equations do not speak clearly to you. Jacob Bronowski once described the verbal argument in mathematical writing as analogous to melody in music, and thus a source of satisfaction in itself. He likened the equations to musical accompaniment that becomes more satisfying with repeated listening. If you follow and retain the melody of arguments and illustrations in Chapters 4 through 7, then the last chapters and their take-home lessons should be useful to you.

I aim further to enable you to read primary journal articles that announce and present new protein structures, including the arcane sections on experimental methods. In most scientific papers, experimental sections are directed primarily toward those who might use the same methods. In crystallographic papers, however, methods sections contain information from which the quality of the model can be roughly judged. This judgement should affect your decision about whether to obtain the model and use it, and whether it is good enough to serve as a guide in drawing the kinds of conclusions you hope to draw. In Chapter 8, to review many concepts, as well as to exercise your new skills, I look at and interpret experimental details in literature reports of a recent structure determination.

Finally, I hope you read this book for pleasure—the sheer pleasure of turning the formerly incomprehensible into the familiar. In a sense, I am attempting to share with you my own pleasure of the past ten years, after my mid-career decision to set aside other interests and finally see how crystallographers produce the molecular models that have been the greatest delight of my teaching. Among those I should thank for opening their labs and giving their time to an old dog trying to learn new tricks are Professors Leonard J. Banaszak, Jens Birktoft, Jeffry Bolin, John Johnson, and Michael Rossman.

I would never have completed this book without the patience of my wife, Pam, who allowed me turn part of our home into a miniature publishing company, nor without the generosity of my faculty colleagues, who allowed me a sabbatical leave during times of great economic stress at the University of Southern Maine. Many thanks to Lorraine Lica, my Acquisitions Editor at Academic Press, who grasped the spirit of this little project from the very beginning and then held me and a full corps of editors, designers, and production workers accountable to that spirit throughout.

Gale Rhodes
Portland, Maine
August 1992

Phase

These still days after frost have let down
the maple leaves in a straight compression
to the grass, a slight wobble from circular to

the east, as if sometime, probably at night, the
wind's moved that way — surely, nothing else
could have done it, really eliminating the *as*

if, although the *as if* can nearly stay since
the wind may have been a big, slow
one, imperceptible, but still angling

off the perpendicular the leaves' fall:
anyway, there was the green-ribbed, yellow,
flat-open reduction: I just now bagged it up.

A. R. Ammons[1]

1 Model and Molecule

Proteins perform many functions in living organisms. For example, some proteins regulate the expression of genes. One class of gene-regulating proteins contains structures known as *zinc fingers*, which bind directly to DNA. Plate 1 shows a complex composed of a double-stranded DNA molecule and three zinc fingers from the mouse protein Zif268.

The protein backbone is shown as a yellow ribbon. The two DNA strands are red and blue. Zinc atoms, which are complexed to side chains in the protein, are green. The green dotted lines near the top center indicate two hydrogen bonds in which nitrogen atoms of arginine-18 (in the protein) share hydrogen atoms with nitrogen and oxygen atoms of guanine-10 (in the DNA), an interaction that holds the sharing atoms about 2.8 Å apart. Studying this complex with modern graphics software, you could zoom in and measure the hydrogen-bond lengths, and find them to be 2.79 and 2.67 Å. You would also learn that all of the protein–DNA interactions are between protein side chains and DNA bases; the protein backbone does not come in contact with the DNA. You could go on to discover all the specific interactions between side chains of Zif268 and base pairs of DNA. You could enumerate the additional hydrogen bonds and other contacts that stabilize this complex and cause Zif268 to recognize a specific sequence of bases in DNA. You might gain some testable insights into how the protein finds the correct DNA sequence amid the vast amount of DNA in the nucleus of a cell. The structure might also lead you to speculate on how alterations in the sequence of amino acids in the protein might result in affinity for different DNA sequences, and thus start you thinking about how to design other DNA-binding proteins.

Now look again at the preceding paragraph and examine its language rather than its content. The language is typical of that in common use to describe molecular structure and interactions as revealed by various experimental methods, including single-crystal X-ray crystallography, the primary subject of this book. In fact, this language is shorthand for more precise but cumbersome statements of what we learn from structural studies. First, Plate 1 of course shows not molecules, but *models* of molecules, in which structures and interactions are *depicted*, not shown. Second, in this specific case, the models are of molecules not in solution, but in the crystalline state, because the models are derived from analysis of X-ray diffraction by crystals of the Zif268/DNA complex. As such, these models depict the average structure of somewhere between 10^{13} and 10^{15} complexes throughout the crystals that were studied. In addition, the structures are averaged over the time of the X-ray experiment, which may be as much as several days.

To draw the conclusions found in the first paragraph requires bringing additional knowledge to bear upon the graphics image, including knowledge of just what we learn from X-ray analysis. (The same could be said for structural models derived from spectroscopic data or any other method.) In short, the graphics image itself is incomplete. It does not reveal things we may know about the complex from other types of experiments, and it does not even reveal all that we learn from X-ray crystallography.

For example, how accurately are the relative positions of atoms known? Are the hydrogen bonds precisely 2.79 and 2.67 Å long, or is there some tolerance in those figures? Is the tolerance large enough to jeopardize the conclusion that a hydrogen bond joins these atoms? Further, do we know anything about how rigid this complex is? Do parts of these molecules vibrate, or do they move with respect to each other? Still further, in the aqueous medium of the cell, does this complex have the same structure as in the crystal, which is a solid? As we examine this model, are we really gaining insight into cellular processes? A final question may surprise you: Does the model fully account for the chemical composition of the crystal? In other words, are any of the known contents of the crystal missing from the model?

The answers to these questions are not revealed in the graphics image, which is more akin to a cartoon than to a molecule. Actually, the answers vary from one model to the next, but they are usually available to the user of crystallographic models. Some of the answers come from X-ray crystallography itself, so the crystallographer does not miss or overlook them. They are simply less accessible to the noncrystallographer than is the graphics image.

Molecular models obtained from crystallography are in wide use as tools for revealing molecular details of life processes. Scientists use models to learn how molecules "work": how enzymes catalyze metabolic reactions, how transport proteins load and unload their molecular cargo, how antibodies bind and destroy foreign substances, and how proteins bind to DNA, perhaps turning genes on and off. It is easy for the user of crystallographic models, being anxious to turn otherwise puzzling information into a mechanism of action, to treat models as everyday objects seen as we see clouds, birds, and trees. But the informed user of models sees more than the graphics image, recognizing it as a static depiction of dynamic objects, as the average of many similar structures, as perhaps lacking parts that are present in the crystal but not revealed by the X-ray analysis, and finally as a fallible interpretation of data. The informed user knows that the crystallographic model is richer than the cartoon.

In the following chapters, I offer you the opportunity to become an informed user of crystallographic models. Knowing the richness and limitations of models requires an understanding of the relationship between data and structure. In Chapter 2, I give an overview of this relationship. In Chapters 3 through 7, I simply expand Chapter 2 in enough detail to produce an intact chain of logic stretching from diffraction data to final model. Topics come in roughly the same order as the tasks that face a crystallographer pursuing an important structure.

As a practical matter, informed use of a model requires reading the crystallographic papers and data files that report the new structure and extracting from them criteria of model quality. In Chapter 8, I discuss these criteria and provide a guided exercise in extracting them. The exercise takes the form of annotated excerpts from a published structure determination and its supporting data. Equipped with the background of previous chapters and experienced with the real-world exercise of a guided tour through a recent publication, you should be able to read new structure publications in the scientific literature and understand how the structures were obtained and be aware of just what is known—and what is still unknown—about the molecules under study.

Chapter 9, "Other Diffraction Methods," builds upon your understanding of X-ray crystallography to help you understand other methods in which diffraction provides insights into the structure of large molecules. These methods include fiber diffraction, neutron diffraction, electron diffraction, and various forms of X-ray spectroscopy. These methods often seem very obscure, but their underlying principles are similar to those of X-ray crystallography.

In Chapter 10, "Other Types of Models," I discuss alternative methods of structure determination: NMR spectroscopy and various forms of theoretical modeling. Just like crystallographic models, NMR and theoretical models are sometimes more, sometimes less, than meets the eye. A brief description of how these models are obtained, along with some analogies among criteria of quality for various types of models, can help make you a wiser user of all types of models.

For new or would-be users of models, I present in Chapter 11 an introduction to molecular modeling, demonstrating how modern graphics programs allow users to display and manipulate models and to perform powerful structure analysis, even on desktop computers. This chapter also provides information on how to use the World Wide Web to obtain graphics programs and learn how to use them. It also provides an introduction to the Protein Data Bank (PDB), a World Wide Web resource from which you can obtain most of the available macromolecular models.

There is an additional, brief chapter that does not lie between the covers of this book. It is the Crystallography Made Crystal Clear (CMCC) Home Page on the World Wide Web at www.usm.maine.edu/~rhodes/CMCC. This web page is devoted to making sure that you can find all the Internet resources mentioned here. Because many Internet resources and addresses change rapidly, I did not include them in these pages; but instead, I refer you to the CMCC Home Page. At that web address, I maintain links to all resources mentioned here or, if they disappear or change markedly, to new ones that serve the same or similar functions. For easy reference, the address of the CMCC Home Page is shown on the cover and title page of this book.

Today's scientific textbooks and journals are filled with stories about the molecular processes of life. The central character in these stories is often a protein or nucleic acid molecule, a thing never seen in action, never perceived directly. We see model molecules in books and on computer screens, and we tend to treat them as everyday objects accessible to our normal perceptions. In fact, models are hard-won products of technically difficult data collection and powerful but subtle data analysis. This book is concerned with where our models of structure come from and how to use them wisely.

2 An Overview of Protein Crystallography

I. Introduction

The most common experimental means of obtaining a detailed picture of a large molecule, allowing the resolution of individual atoms, is to interpret the diffraction of X rays from many identical molecules in an ordered array like a crystal. This method is called *single-crystal X-ray crystallography.* As of this writing, roughly 8000 protein and nucleic-acid structures have been obtained by this method. In addition, the structures of roughly 1300 macromolecules, mostly proteins of fewer than 150 residues, have been solved by nuclear magnetic resonance (NMR) spectroscopy, which provides a model of the molecule in solution, rather than in the crystalline state. Finally, there are theoretical models, built by analogy with the structures of known proteins having similar sequence, or based on simulations of protein folding. All methods have their strengths and weaknesses, and they will undoubtedly coexist as complementary methods for the foreseeable future. One of the goals of this book is to make users of crystallographic models aware of the strengths and weaknesses of X-ray crystallography, so that users' expectations of the resulting models are in keeping with the limitations of crystallographic methods. Chapter 10 provides, in brief, complementary information about other types of models.

This chapter provides a simplified overview of how researchers use the technique of X-ray crystallography to learn macromolecular structures. Chapters 3–8 are simply expansions of the material in this chapter. In order to keep the language simple, I will speak primarily of proteins, but the concepts I describe apply to all macromolecules and macromolecular assemblies that possess ordered structure, including carbohydrates, nucleic acids, and nucleoprotein complexes like ribosomes and whole viruses.

A. Obtaining an image of a microscopic object

When we see an object, light rays bounce off (are diffracted by) the object and enter the eye through the lens, which reconstructs an image of the object and focuses it on the retina. In a simple microscope, an illuminated object is placed just beyond one focal point of a lens, which is called the *objective* lens. The lens collects light diffracted from the object and reconstructs an image beyond the focal point on the opposite side of the lens, as shown in Fig. 2.1.

For a simple lens, the relationship of object position to image position in Fig. 2.1 is $(OF)(IF') = (FL)(F'L)$. Because the distances FL and $F'L$ are constants (but not necessarily equal) for a fixed lens, the distance OF is inversely proportional to the distance IF'. Placing the object near the focal point

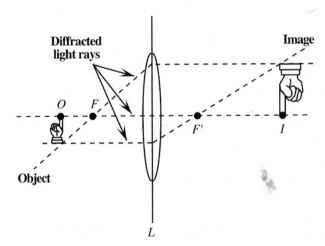

Figure 2.1 Action of a simple lens. Rays parallel to the lens strike the lens and are refracted into paths passing through a focus. Rays passing through a focus strike the lens and are refracted into paths parallel to the lens axis. As a result, the lens produces an image at I of an object at O, such that $(OF)(IF') = (FL)(F'L)$.

F results in a magnified image produced at a considerable distance from F' on the other side of the the lens, which is convenient for viewing. In a compound microscope, the most common type, an additional lens, the *eyepiece*, is added to magnify the image produced by the objective lens.

B. Obtaining images of molecules

In order for the object to diffract light and thus be visible under magnification, the wavelength (λ) of the light must be, roughly speaking, no larger than the object. Visible light, which is electromagnetic radiation with wavelengths of 400–700 nm (nm = 10^{-9} m), cannot produce an image of individual atoms in protein molecules, in which bonded atoms are only about 0.15 nm or 1.5 Å ($\text{Å} = 10^{-10}$ m) apart. Electromagnetic radiation of this wavelength falls into the X-ray range, so X rays are diffracted by even the smallest molecules. X-ray analysis of proteins seldom resolves the hydrogen atoms, so the protein models described in this book include elements on only the second and higher rows of the periodic table. The positions of all hydrogen atoms can be deduced on the assumption that bond lengths, bond angles, and conformational angles in proteins are just like those in small organic molecules.

Even though individual atoms diffract X rays, it is still not possible to produce a focused image of a molecule, for two reasons. First, X rays cannot be focused by lenses. Crystallographers sidestep this problem by measuring the directions and strengths (intensities) of the diffracted X rays and then using a computer to simulate an image-reconstructing lens. In short, the computer acts as the lens, computing the image of the object and then displaying it on a screen or drawing it on paper (Fig. 2.2).

Second, a single molecule is a very weak scatterer of X rays. Most of the X rays will pass through a single molecule without being diffracted, so the diffracted beams are too weak to be detected. Analyzing diffraction from crystals, rather than individual molecules, solves this problem. A crystal of a protein contains many ordered molecules in identical orientations, so each molecule diffracts identically, and the diffracted beams for all molecules augment each other to produce strong, detectable X-ray beams.

C. A thumbnail sketch of protein crystallography

In brief, determining the structure of a protein by X-ray crystallography entails growing high-quality crystals of the purified protein, measuring the directions and intensities of X-ray beams diffracted from the crystals, and using a computer to simulate the effects of an objective lens and thus produce an

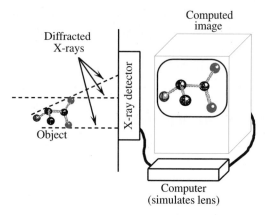

Figure 2.2　Crystallographic analogy of lens action. X-rays diffracted from the object are received and measured by a detector. The measurements are fed to a computer, which simulates the action of a lens to produce a graphics image of the object.

image of the crystal's contents, like the small section of a molecular image shown in Plate 2*a*. Finally, that image must be interpreted, which entails displaying it by computer graphics and building a molecular model that is consistent with the image (Plate 2*b*).

The resulting model is often the only product of crystallography that the user sees. It is therefore easy to think of the model as a real entity that has been directly observed. In fact, our "view" of the molecule is quite indirect. Understanding just how the crystallographer obtains models of protein molecules from diffraction measurements is essential to fully understanding how to use models properly.

II. Crystals

A. *The nature of crystals*

Under certain circumstances, many molecular substances, including proteins, solidify to form crystals. In entering the crystalline state from solution, individual molecules of the substance adopt one or a few identical orientations. The resulting crystal is an orderly three-dimensional array of molecules, held together by noncovalent interactions. Figure 2.3 shows such a crystalline array of molecules.

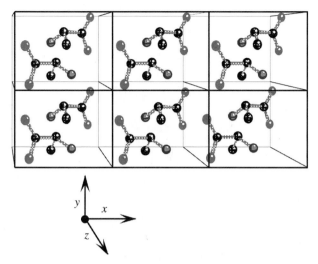

Figure 2.3 Six unit cells in a crystalline lattice. Each unit cell contains two molecules of alanine (hydrogen atoms not shown) in different orientations.

The lines in the figure divide the crystal into identical *unit cells*. The array of points at the corners or vertices of unit cells is called the *lattice*. The unit cell is the smallest and simplest volume element that is completely representative of the whole crystal. If we know the exact contents of the unit cell, we can imagine the whole crystal as an efficiently packed array of many unit cells stacked beside and on top of each other, more or less like identical boxes in a warehouse.

From crystallography, we obtain an image of the electron clouds that surround the molecules in the average unit cell in the crystal. We hope this image will allow us to locate all atoms in the unit cell. The location of an atom is usually given by a set of three-dimensional Cartesian coordinates, x, y, and z. One of the vertices (a lattice point or any other convenient point) is used as the origin of the unit cell's coordinate system and is assigned the coordinates $x = 0$, $y = 0$, and $z = 0$, usually written (0,0,0). See Fig. 2.4.

B. Growing crystals

Crystallographers grow crystals of proteins by slow, controlled precipitation from aqueous solution under conditions that do not denature the protein. A number of substances cause proteins to precipitate. Ionic compounds (salts) precipitate proteins by a process called "salting out." Organic solvents also cause precipitation, but they often interact with hydrophobic

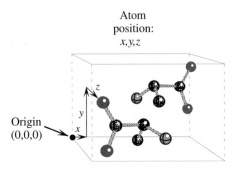

Figure 2.4   One unit cell from Fig. 2.3. The position of an atom in the unit cell can be specified by a set of spatial coordinates *x, y, z*.

portions of proteins and thereby denature them. The water-soluble polymer polyethylene glycol (PEG) is widely used because it is a powerful precipitant and a weak denaturant. It is available in preparations of different average molecular masses, such as PEG 400, with average molecular mass of 400 daltons.

One simple means of causing slow precipitation is to add denaturant to an aqueous solution of protein until the denaturant concentration is just below that required to precipitate the protein. Then water is allowed to evaporate slowly, which gently raises the concentration of both protein and denaturant until precipitation occurs. Whether the protein forms crystals or instead forms a useless amorphous solid depends on many properties of the solution, including protein concentration, temperature, pH, and ionic strength. Finding the exact conditions to produce good crystals of a specific protein often requires many careful trials and is perhaps more art than science. I will examine crystallization methods in Chapter 3.

III. Collecting X-ray data

Figure 2.5 depicts the collection of X-ray diffraction data. A crystal is mounted between an X-ray source and an X-ray detector. The crystal lies in the path of a narrow beam of X rays coming from the source. A simple detector is X-ray film, which when developed exhibits dark spots where X-ray beams have impinged. These spots are called *reflections* because they emerge from the crystal as if reflected from planes of atoms.

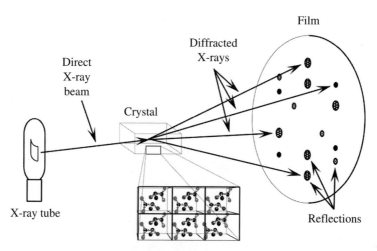

Figure 2.5 Crystallographic data collection. The crystal diffracts the source beam into many discrete beams, each of which produces a distinct spot (reflection) on the film. The positions and intensities of these reflections contain the information needed to determine molecular structures.

Figure 2.6 shows the complex diffraction pattern of X rays produced on film by a protein crystal. Notice that the crystal diffracts the source beam into many discrete beams, each of which produces a distinct reflection on the film. The greater the intensity of the X-ray beam that reaches a particular position, the darker the reflection.

An optical scanner precisely measures the position and the intensity of each reflection and transmits this information in digital form to a computer for analysis. The position of a reflection can be used to obtain the direction in which that particular beam was diffracted by the crystal. The intensity of a reflection is obtained by measuring the optical absorbance of the spot on the film, giving a measure of the strength of the diffracted beam that produced the spot. The computer program that reconstructs an image of the molecules in the unit cell requires these two parameters, the beam intensity and direction, for each diffracted beam.

Although film for data collection has largely been replaced by devices that feed diffraction data (positions and intensities of each reflection) directly into computers, I will continue to speak of the data as if collected on film because of the simplicity of that format, and because diffraction patterns are usually published in a form identical to their appearance on film. I will discuss other methods of collecting data in Chapter 4.

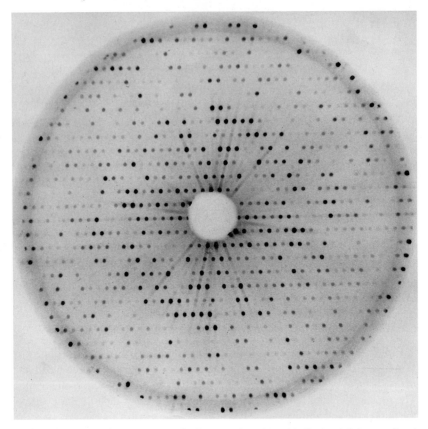

Figure 2.6 Diffraction pattern from a crystal of the MoFe (molybdenum–iron) protein of the enzyme nitrogenase from *Clostridium pasteurianum*. Notice that the reflections lie in a regular pattern, but their intensities (darkness of spots) are highly variable. [The hole in the middle of the pattern results from a small metal disk (beam stop) used to prevent the direct X-ray beam, most of which passes straight through the crystal, from destroying the center of the film.] Photo courtesy of Professor Jeffery Bolin.

IV. Diffraction

A. Simple objects

You can develop some visual intuition for the information available from X-ray diffraction by examining the diffraction patterns of simple objects like spheres or arrays of spheres (Figs. 2.7–2.10). Figure 2.7 depicts diffraction by a single sphere, shown in cross section on the left. The diffraction pattern, on

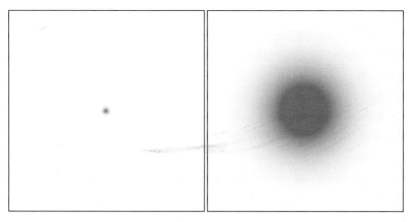

Figure 2.7   Sphere (cross-section, on left) and its diffraction pattern (right). Images for Figures 2.7–2.10 were generously provided by Dr. Kevin Cowtan.

the right, exhibits high intensity at the center, and smoothly decreasing intensity as the diffraction angle increases.[1]

For now, just accept the fact that diffraction by a sphere produces this pattern, and think of it as the diffraction signature of a sphere. In a sense, you are already equipped to do very simple structure determination; that is, you can now recognize a simple sphere by its diffraction pattern.

B. Arrays of simple objects: Real and reciprocal lattices

Figure 2.8 depicts diffraction by a crystalline array of spheres, with a cross section of the crystal on the left, and its diffraction pattern on the right.

The diffraction pattern, like that produced by crystalline nitrogenase (Fig. 2.6), consists of reflections (spots) in an orderly array on the film. The spacing of the reflections varies with the spacing of the spheres in their array. Specifically, observe that although the lattice spacing of the crystal is smaller vertically, the diffraction spacing is smaller horizontally. In fact, there is a simple inverse relationship between the spacing of unit cells in the crystalline lattice, called the *real lattice*, and the spacing of reflections in the lattice on the film, which, because of its inverse relationship to the real lattice, is called the *reciprocal lattice*.

---

[1]The images shown in Figures 2.7–2.10 are computed, rather than experimental, diffraction patterns. Computation of these patterns involves use of the Fourier transform (Section V.E).

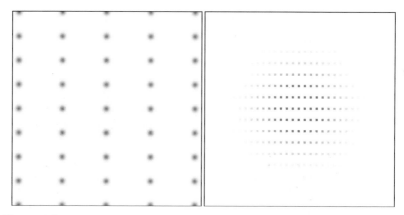

Figure 2.8 Lattice of spheres (left) and its diffraction pattern (right). If you look at the pattern and blur your eyes, you will see the diffraction pattern of a sphere. The pattern is that of the average sphere in the real lattice, but it is sampled at the reciprocal lattice points.

Because the real lattice spacing is inversely proportional to the spacing of reflections, crystallographers can calculate the dimensions, in angstroms, of the unit cell of the crystalline material from the spacings of the reciprocal lattice on the X-ray film (Chapter 4). The simplicity of this relationship is a dramatic example of how the macroscopic dimensions of the diffraction pattern are connected to the submicroscopic dimensions of the crystal.

C. Intensities of reflections

Now look at the intensities of the reflections in Fig. 2.8. Some are intense ("bright"), whereas others are weak or perhaps missing from the otherwise evenly spaced pattern. These variations in intensity contain important information. If you blur your eyes slightly while looking at the diffraction pattern, so that you cannot see individual spots, you will see the intensity pattern characteristic of diffraction by a sphere, with lower intensities farther from the center, as in Fig. 2.7. (You just determined your first crystallographic structure.) The diffraction pattern of spheres in a lattice is simply the diffraction pattern of the average sphere in the lattice, but this pattern is incomplete. The pattern is *sampled* at points whose spacings vary inversely with real-lattice spacings. The pattern of varied intensities is that of the *average* sphere because all the spheres contribute to the observed pattern. To put it another way, the observed pattern of intensities is actually a superposition of the many identical diffraction patterns of all the spheres.

D. Arrays of complex objects

This relationship between (1) diffraction by a single object and (2) diffraction by many identical objects in a lattice holds true for complex objects also. Figure 2.9 depicts diffraction by six spheres that form a planar hexagon, like the six carbons in benzene.

Notice the starlike six-fold symmetry of the diffraction pattern. Again, just accept this pattern as the diffraction signature of a hexagon of spheres. (Now you know enough to recognize *two* simple objects by their diffraction patterns.) Figure 2.10 depicts diffraction by these hexagonal objects in a lattice of the same dimensions as that in Fig. 2.8.

As before, the spacing of reflections varies reciprocally with lattice spacing, but if you blur your eyes slightly, or compare Figs. 2.9 and 2.10 carefully, you will see that the starlike signature of a single hexagonal cluster is present in Fig. 2.10. From these simple examples, you can see that the reciprocal-lattice spacing (the spacing of reflections in the diffraction pattern) is characteristic of (inversely related to) the spacing of identical objects in the crystal, whereas the reflection intensities are characteristic of the shape of the individual objects. From the reciprocal-lattice spacing in a diffraction pattern, we can compute the dimensions of the unit cell. From the intensities of the reflections,

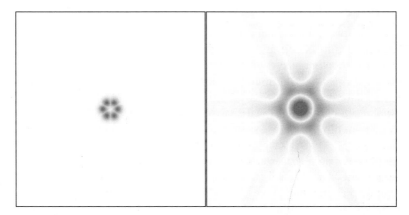

Figure 2.9 A planar hexagon of spheres (left) and its diffraction pattern (right).

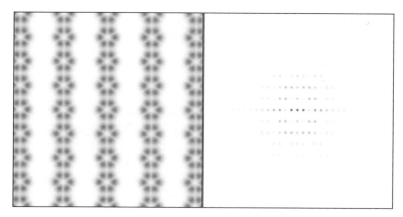

Figure 2.10   Lattices of hexagons (left) and its diffraction pattern (right). If you look at the pattern and blur your eyes, you will see the diffraction pattern of a hexagon. The pattern is that of the average hexagon in the real lattice, but it is sampled at the reciprocal lattice points.

we can learn the shape of the individual molecules that compose the crystal. It is actually advantageous that the object's diffraction pattern is sampled at reciprocal-lattice positions. This sampling reduces the number of intensity measurements we must take from the film and makes it easier to program a computer to locate and measure the intensities.

E. Three-dimensional arrays

Unlike the two-dimensional arrays in these examples, a crystal is a three-dimensional array of objects. If we rotate the crystal in the X-ray beam, a different cross section of objects will lie perpendicular to the beam, and we will see a different diffraction pattern. In fact, just as the two-dimensional arrays of objects we have discussed are cross sections of objects in the three-dimensional crystal, each two-dimensional array of reflections (each diffraction pattern recorded on film) is a cross section of a three-dimensional lattice of reflections. Figure 2.11 shows a hypothetical three-dimensional diffraction pattern, with the reflections that would be produced by all possible orientations of a crystal in the X-ray beam.

Notice that only one plane of the three-dimensional diffraction pattern is superimposed on the film. With the crystal in the orientation shown, reflections shown in the plane of the film (solid spots) are the only reflections that produce spots on the film. In order to measure the directions and intensities of

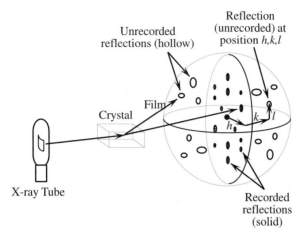

Figure 2.11   Crystallographic data collection, showing reflections measured at one particular crystal orientation (solid, on film) and those that could be measured at other orientations (hollow, within the sphere but not on the film). The relationship between measured and unmeasured reflections is more complex than shown here (see Chapter 4).

all additional reflections (shown as hollow spots), the crystallographer must collect diffraction patterns from all unique orientations of the crystal with re-spect to the X-ray beam. The direct result of crystallographic data collection is a list of intensities for each point in the three-dimensional reciprocal lattice. This set of data is the raw material for determining the structures of molecules in the crystal.

(*Note*: The spatial relationship involving beam, crystal, film, and reflections is more complex than shown here. I will discuss the actual relationship in Chapter 4.)

V. Coordinate systems in crystallography

Each reflection can be assigned three coordinates or *indices* in the imaginary three-dimensional space of the diffraction pattern. This space, the strange land where the reflections live, is called *reciprocal space*. Crystallographers usually use h, k, and l to designate the position of an individual reflection in the recip-rocal space of the diffraction pattern. The central reflection (the round solid spot at the center of the film in Fig. 2.11) is taken as the origin in reciprocal space and assigned the coordinates $(h,k,l) = (0,0,0)$, usually written $hkl = 000$. (The 000 reflection is not measurable because it is always obscured by X rays that pass straight through the crystal.) The other reflections are assigned whole-number coordinates counted from this origin, so the indices h, k, and l are integers. Thus the parameters we can measure and analyze in the X-ray diffraction pattern are the position hkl and the intensity I_{hkl} of each reflection. The position of a reflection is related to the angle by which the diffracted beam diverges from the source beam. For a unit cell of known dimensions, the angle of divergence uniquely specifies the indices of a reflection (see Chapter 4).

Alternatively, actual distances, rather than reflection indices, can be measured in reciprocal space. Because the dimensions of reciprocal space are the inverse of dimensions in the real space of the crystal, distances in reciprocal space are expressed in the units Å^{-1} (called *reciprocal angstroms*). Roughly speaking, the inverse of the reciprocal-space distance from the origin out to the most distant measurable reflections gives the potential resolution of the model that we can obtain from the data. So a crystal that gives measurable reflections out to a distance of $1/(3\ \text{Å})$ from the origin should yield a model with a resolution of 3 Å.

The crystallographer works back and forth between two different coordinate systems. I will review them briefly. The first system (see Fig. 2.4) is the unit cell (real space), where an atom's position is described by its coordinates x,y,z.

Figure 2.12   Fun in reciprocal space. © The New Yorker Collection, 1991. John O'Brien, from cartoonbank.com. All rights reserved.

A vertex of the unit cell, or any other convenient position, is taken as the origin, with coordinates $x,y,z = (0,0,0)$. Coordinates in real space designate real spatial positions within the unit cell. Real-space coordinates are usually given in angstroms or nanometers, or in fractions of unit cell dimensions. The second system (see Fig. 2.11) is the three-dimensional diffraction pattern (reciprocal space), where a reflection's position is described by its indices hkl. The central reflection is taken as the origin with the index 000 (round black dot at center of sphere). The position of a reflection is designated by counting reflections from 000, so the indices h, k, and l are integers. Distances in reciprocal space, expressed in reciprocal angstroms or reciprocal nanometers, are used to judge the potential resolution that the diffraction data can yield.

Like Alice's looking-glass world, reciprocal space may seem strange to you at first (Fig. 2.12). We will see, however, that some aspects of crystallography are actually easier to understand, and some calculations are more convenient, in reciprocal space than in real space (Chapter 4).

VI. The mathematics of crystallography: A brief description

The problem of determining the structure of objects in a crystalline array from their diffraction pattern is, in essence, a matter of converting the experimentally accessible information in the reciprocal space of the diffraction pattern to otherwise inaccessible information about the real space inside the unit cell. Remember that a computer program that makes this conversion is acting as a simulated lens to reconstruct an image from diffracted radiation. Each reflection is produced by a beam of electromagnetic radiation (X rays), so the computations entail treating the reflections as waves and recombining these waves to produce an image of the molecules in the unit cell.

A. Wave equations: Periodic functions

Each reflection is the result of diffraction from complicated objects, the molecules in the unit cell, so the resulting wave is complicated also. Before considering how the computer represents such an intricate wave, let us consider mathematical descriptions of the simplest waves.

A simple wave, like that of visible light or X rays, can be described by a periodic function, for instance, an equation of the form

$$f(x) = F \cos 2\pi (hx + \alpha) \tag{2.1}$$

or

$$f(x) = F \sin 2\pi (hx + \alpha). \tag{2.2}$$

In these functions, $f(x)$ specifies the vertical height of the wave at any horizontal position x along the wave. The variable x and the constant α are angles expressed in fractions of the wavelength; that is, $x = 1$ implies a position of one full wavelength (2π radians or 360°) from the origin. The constant F specifies the amplitude (the height of the crests and troughs) of the wave. For example, the crests of the wave $f(x) = 3 \cos 2\pi x$ are three times as high and the troughs are three times as deep as those of the wave $f(x) = \cos 2\pi x$ (compare b with a in Fig. 2.13).

The constant h in a simple wave equation specifies the frequency or wavelength of the wave. For example, the wave $f(x) = \cos 2\pi (5x)$ has five times the frequency (or one-fifth the wavelength) of the wave $f(x) = \cos 2\pi x$ (compare c with a in Fig. 2.13). (In the wave equations used in this book, h takes on integral values only.)

Finally, the constant α specifies the phase of the wave, that is, the position of the wave with respect to the origin of the coordinate system on which the wave is plotted. For example, the position of the wave $f(x) = \cos 2\pi (x + 1/4)$ is shifted by one-quarter of 2π radians (or one-quarter of a wavelength, or 90°) from the position of the wave $f(x) = \cos 2\pi x$ (compare Fig. 2.13d with Fig. 2.13a). Because the wave is repetitive, with a repeat distance of one wavelength or 2π radians, a phase of 1/4 is the same as a phase of $1\,1/4$, or $2\,1/4$, or $3\,1/4$, and so on. In radians, a phase of 0 is the same as a phase of 2π, or 4π, or 6π, and so on.

These equations describe one-dimensional waves, in which a property (in this case, the height of the wave) varies is one direction. Visualizing a one-dimensional function $f(x)$ requires a two-dimensional graph, with the second dimension used to represent the numerical value of $f(x)$. For example, if $f(x)$ describes the electrical part of an electromagnetic wave, the x-axis is the direction the wave is moving, and the height of the wave at any position on the x-axis represents the momentary strength of the electrical field at a distance x from the origin. The field strength is in no real sense perpendicular to x, but it is convenient to use the perpendicular direction to show the numerical value of the field strength. In general, visualizing a function in n dimensions requires $n + 1$ dimensions.

B. Complicated periodic functions: Fourier series

As discussed in Section VI.A, any simple sine or cosine wave can therefore be described by three constants—the amplitude F, the frequency h, and the

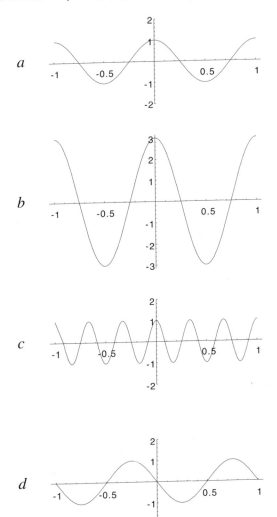

Figure 2.13 Graphs of four simple wave equations $f(x) = F\cos 2\pi(hx + \alpha)$. (a) $F = 1, h = 1, \alpha = 0$: $f(x) = \cos 2\pi(x)$. (b) $F = 3, h = 1, \alpha = 0$: $f(x) = 3\cos 2\pi(x)$. Increasing F increases the amplitude of the wave. (c) $F = 1, h = 3, \alpha = 0$: $f(x) = \cos 2\pi(3x)$. Increasing h increases the frequency (or decreases the wavelength λ) of the wave. (d) $F = 1, h = 1, \alpha = 1/4$: $f(x) = \cos 2\pi(x + 1/4)$. Changing α changes the phase (position) of the wave.

phase α. It is less obvious that far more complicated waves can also be described with this same simplicity. The French mathematician Jean Baptiste Joseph Fourier (1768–1830) showed that even the most intricate periodic functions can be described as the sum of simple sine and cosine functions whose wavelengths are integral fractions of the wavelength of the complicated function. Such a sum is called a *Fourier series* and each simple sine or cosine function in the sum is called a *Fourier term*.

Figure 2.14 shows a periodic function, called a step function, and the beginning of a Fourier series that describes it. A method called *Fourier synthesis* is used to compute the sine and cosine terms that describe a complex wave, which I will call the "target" of the synthesis. I will discuss the results of Fourier synthesis, but not the method itself. In the example of Fig. 2.14, the first four terms produced by Fourier synthesis are shown individually (f_0 through f_3), and each is added sequentially to the Fourier series. Notice that the first term in the series, $f_0 = 1$, simply displaces the sums upward so that they have only positive values like the target function. (Sine and cosine functions themselves have both positive and negative values, with average values of zero.) The second term $f_1 = \cos 2\pi x$, has the same wavelength as the step function, and wavelengths of subsequent terms are simple fractions of that wavelength. (It is equivalent to say, and it is plain in the equations, that the frequencies h are simple multiples of the frequency of the step function.) Notice that the sum of only the first few Fourier terms merely approximates the target. If additional terms of shorter wavelength are computed and added, the fit of the approximated wave to the target improves, as shown by the sum of the first six terms. Indeed, using the tenets of Fourier theory, it can be proved that such approximations can be made as similar as desired to the target waveform, simply by including enough terms in the series.

Look again at the components of the Fourier series, functions f_0 through f_3. The low-frequency terms like f_1 approximate the gross features of the target wave. Higher-frequency terms like f_3 improve the approximation by filling in finer details, for example, making the approximation better in the sharp corners of the target function.

Figure 2.14 Beginning of a Fourier series to approximate a target function, in this case, a step function or square wave. $f_0 = 1; f_1 = \cos 2\pi(x); f_2 = (-1/3) \cos 2\pi(3x); f_3 = 1/5) \cos 2\pi(5x)$. In the left column are the target and terms f_1 through f_3. In the right column are f_0 and the succeeding sums as each term is added to f_0. Notice that the approximaton improves (i.e. each successive sum looks more like the target) as the number of Fourier terms in the sum increase. In the last graph, terms f_4, f_5 and f_6 are added (but not shown separately) to show further improvement in the approximation.

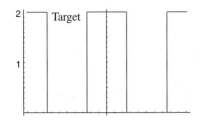

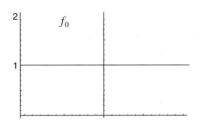

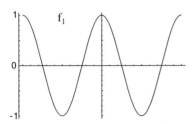

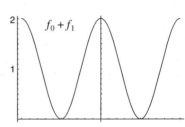

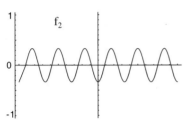

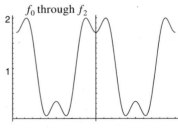

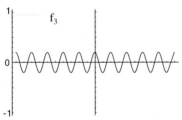

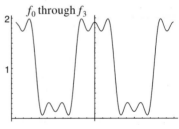

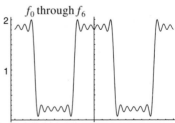

C. Structure factors: Wave descriptions of X-ray reflections

Each diffracted X ray that arrives at the film to produce a recorded reflection can also be described as the sum of the contributions of all scatterers in the unit cell. The sum that describes a diffracted ray is called a *structure–factor equation*. The computed sum for the reflection *hkl* is called the *structure factor* F_{hkl}. As I will show in Chapter 4, the structure–factor equation can be written in several different ways. For example, one useful form is a sum in which each term describes diffraction by one atom in the unit cell, and thus the series contains the same number of terms as the number of atoms.

If diffraction by atom *A* in Fig. 2.15 is represented by f_A, then one diffracted ray (producing one reflection) from the unit cell of Fig. 2.15 is described by a structure–factor equation of this form:

$$F_{hkl} = f_A + f_B + \ldots + f_{A'} + f_{B'} = \ldots + f_{F'}. \tag{2.3}$$

The structure–factor equation implies, and correctly so, that each reflection on the film is the result of diffractive contributions from all atoms in the unit cell. That is, every atom in the unit cell contributes to every reflection in the diffraction pattern. The structure factor is a wave created by the superposition of many individual waves, each resulting from diffraction by an individual atom.

D. Electron-density maps

To be more precise about diffraction, when we direct an X-ray beam toward a crystal, the actual diffractors of the X rays are the clouds of electrons in the molecules of the crystal. Diffraction should therefore reveal the distribution of electrons, or the electron density, of the molecules. Electron density, of course, reflects the molecule's shape; in fact, you can think of the molecule's boundary as a van der Waals surface, the surface of a cloud of electrons that surrounds the molecule. Because, as noted earlier, protein molecules are ordered, and because, in a crystal, the molecules are in an ordered array, the electron density in a crystal can be described mathematically by a periodic function.

If we could walk through the crystal depicted in Fig. 2.3, along a linear path parallel to a cell edge, and carry with us a device for measuring electron density, our device would show us that the electron density varies along our path in a complicated periodic manner, rising as we pass through molecules, falling in the space between molecules, and repeating its variation identically as we pass through each unit cell. Because this statement is true for linear paths parallel to all three cell edges, the electron density, which describes the surface

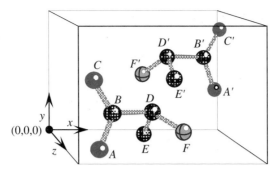

Figure 2.15 Every atom contributes to every reflection in the diffraction pattern, as described for this unit cell by Eq. (2.3).

features and overall shape of all molecules in the unit cell, is a three-dimensional periodic function. I will refer to this function as $\rho(x,y,z)$, implying that it specifies a value ρ for electron density at every position x,y,z in the unit cell. A graph of the function is an image of the electron clouds that surround the molecules in the unit cell. The most readily interpretable graph is a contour map—a drawing of a surface along which there is constant electron density (refer to Plate 2*a*). The graph is called an *electron-density map*. The map is, in essence, a fuzzy image of the molecules in the unit cell. The goal of crystallography is to obtain the mathematical function whose graph is the desired electron-density map.

E. Electron density from structure factors

Because the electron density we seek is a complicated periodic function, it can be described as a Fourier series. Do the many structure–factor equations, each a sum of wave equations describing one reflection in the diffraction pattern, have any connection with the Fourier series that describes the electron density? As mentioned earlier, each structure–factor equation can be written as a sum in which each term describes diffraction from one atom in the unit cell. But this is only one of many ways to write a structure–factor equation. Another way is to imagine dividing the electron density in the unit cell into many small volume elements by inserting planes parallel to the cell edges (Fig. 2.16).

These volume elements can be as small and numerous as desired. Now because the true diffractors are the clouds of electrons, each structure–factor equation can be written as a sum in which each term describes diffraction by the electrons in one volume element. In this sum, each term contains the average numerical value of the desired electron density function $\rho(x,y,z)$ within

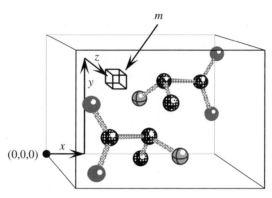

Figure 2.16   Small volume element *m* within the unit cell, one of many elements formed by subdividing the unit cell with planes parallel to the cell edges. The average electron density within *m* is $\rho_m(x,y,z)$. Every volume element contributes to every reflection in the diffraction pattern, as described by Eq. (2.4).

one volume element. If the cell is divided into *n* elements, and the average electron density in volume element *m* is ρ_m, then one diffracted ray from the unit cell of Fig. 2.16 is described by a structure–factor equation of this form:

$$F_{hkl} = f(\rho_1) + f(\rho_2) + \cdots + f(\rho_m) + \cdots + f(\rho_n). \qquad (2.4)$$

So each reflection is described by an equation like this, giving us a large number of equations describing reflections in terms of the electron density. Is there any way to solve these equations for the function $\rho(x,y,z)$ in terms of the measured reflections? After all, structure factors like Eq. (2.4) describe the reflections in terms of $\rho(x,y,z)$, which is precisely the function the crystallographer is trying to learn. I will show in Chapter 5 that a mathematical operation called the Fourier transform solves the structure–factor equations for the desired function $\rho(x,y,z)$, just as if they were a set of simultaneous equations describing $\rho(x,y,z)$ in terms of the amplitudes, frequencies, and phases of the reflections.

The Fourier transform describes precisely the mathematical relationship between an object and its diffraction pattern. In Figs. 2.7–2.10, the diffraction patterns are the Fourier transforms of the corresponding objects or arrays of objects. To put it another way, the Fourier transform is the lens-simulating operation that a computer performs to produce an image of molecules (or more precisely, of electron clouds) in the crystal. This view of $\rho(x,y,z)$ as the Fourier transform of the structure factors implies that if we can measure three parameters—amplitude, frequency, and phase—of *each* reflection, then we can obtain the function $\rho(x,y,z)$, graph the function, and "see" a fuzzy image of the molecules in the unit cell.

F. Electron density from measured reflections

Are all three of these parameters accessible in the data on our films? We will see in Chapter 5 that the measurable intensity I_{hkl} of one reflection gives the amplitude of one Fourier term in the series that describes $\rho(x,y,z)$, and that the position hkl specifies the frequency for that term. But the phase α of each reflection is not recorded on the film. In Chapter 6, we will see how to obtain the phase of each reflection, completing the information we need to calculate $\rho(x,y,z)$.

A final note: Even though we cannot measure phases by simply collecting diffraction patterns, we can compute them from a known structure, and we can depict them by adding color to images like those of Figures 2.7–2.10. In his innovative World Wide Web *Book of Fourier*[2], Kevin Cowtan illustrates phases in diffraction patterns in this clever manner. For example, Plate 3*a* shows a lattice of simple objects, each one like the carbon atoms in ethylbenzene. Plate 3*b* is the computed Fourier transform of (*a*). Image (*c*) depicts a lattice of the objects in (*a*), and *d*) depicts the corresponding diffraction pattern.

Because patterns (*b*) and (*d*) were *computed* from objects of known structure, rather than measured experimentally from real objects, the phases are known. The phase of each reflection is depicted by its color, according to the color wheel (*f*). The phase can be expressed as an angle between 0° and 360° [this is the angle α in Eqs. (2.1) or (2.2)]. In Plate 3, the phase angle of each region (in *b*) or reflection (in *d*) is the angle that corresponds to the angle of its color on the color wheel (*f*). For example, red corresponds to a phase angle of 0°, and green to an angle of about 135°. So a dark red reflection has a high intensity and a phase angle of 0°. A pale green reflection has a low intensity and a phase angle of about 135°. This depiction, then, gives a full description of each reflection, including the phase angle that we do not learn from diffraction experiments, which would give us only the intensities, as shown in (*e*). In a sense then, Figs. 2.7 through 2.10 and Plate 3*e* show *diffraction* patterns, whereas Plates 3*b* and 3*d* show *structure–factor* patterns, which depict the structure factors fully. Note again that (*d*) is a sampling of (*b*) at points corresponding to the reciprocal lattice of the lattice in (*c*). In other words, the diffraction pattern (*d*) still contains the diffraction signature, including both intensities and phases, of the object in (*a*).

In these terms, I will restate a central problem of crystallography: In order to determine a structure, we need a full-color version of the diffraction pattern—that is, a full description of the structure factors. But diffraction experiments give us only the black-and-white version, the intensities of the

---

[2]Access to Kevin Cowtan's *Book of Fourier* is provided at the CMCC Home Page, www.usm.maine.edu/~rhodes/CMCC.

reflections, but no information about their phases. We must learn the phase angles from further experimentation, as described fully in Chapter 6.

G. Obtaining a model

Having obtained $\rho(x,y,z)$, we graph the function to produce an electron-density map, an image of the molecules in the unit cell. Finally, we interpret the map by building a model that fits it (refer to Plate 2b). In interpreting the molecular image and building the model, a crystallographer takes advantage of all current knowledge about the protein under investigation, as well as knowledge about protein structure in general. Probably the most important information required is the sequence of amino acids in the protein. In a few rare instances, the amino-acid sequence has been learned from the crystallographic structure. But in almost all cases, crystallographers know the sequence to start with, from the work of chemists or molecular biologists, and use it to help them interpret the image obtained from crystallography. In effect, the crystallographer starts with knowledge of the chemical structure, but without knowledge of the conformation. Interpreting the image amounts to finding a chemically realistic conformation that fits the image precisely.

A crystallographer interprets a map by displaying it on a graphics computer and building a graphics model within it. The final model must be (1) consistent with the image and (2) chemically realistic; that is, it must possess bond lengths, bond angles, conformational angles, and distances between neighboring groups that are all in keeping with established principles of molecular structure and stereochemistry. With such a model in hand, the crystallographer can begin to explore the model for clues about its function.

In Chapters 3–7, I will take up in more detail the principles introduced in this chapter.

3 Protein Crystals

I. Properties of protein crystals

A. Introduction

As the term *X-ray crystallography* implies, the sample being examined is in the crystalline state. Crystals of many proteins and other biomolecules have been obtained and analyzed in the X-ray beam. A few macromolecular crystals are shown in Fig. 3.1.

In these photographs, the crystals appear much like inorganic materials such as sodium chloride. But there are several important differences between protein crystals and ionic solids.

B. Size, structural integrity, and mosaicity

Whereas inorganic crystals can often be grown to dimensions of several centimeters or larger, it is frequently impossible to grow protein crystals as large as 1 mm in their shortest dimension. In addition, larger crystals are often twinned (two or more crystals grown into each other at different orientations)

or otherwise imperfect and not usable. Roughly speaking, protein crystallography requires a crystal of at least 0.2 mm in its shortest dimension, although modern methods of data collection can sometimes succeed with smaller crystals.

Inorganic crystals derive their structural integrity from the electrostatic attraction of fully charged ions. On the other hand, protein crystals are held together by weaker forces, primarily hydrogen bonds between hydrated protein surfaces. In other words, proteins in the crystal stick to each other primarily by hydrogen bonds through intervening water molecules. Protein crystals are thus much more fragile than inorganic crystals; gentle pressure with a needle is enough to crush the hardiest protein crystal. Growing, handling, and mounting crystals for analysis thus require very gentle techniques. Protein crystals are usually harvested, examined, and mounted for crystallography within their *mother liquor,* the solution in which they formed.

The textbook image of a crystal is that of a perfect array of unit cells stretching throughout. Real macroscopic crystals are actually mosaics of many submicroscopic arrays in rough alignment with each other. The result of mosaicity is that an X-ray reflection actually emerges from the crystal as a narrow cone rather than a perfectly linear beam. Thus the reflection must be measured over a very small range of angles, rather than at a single, well-defined angle. In protein crystals, composed as they are of relatively flexible molecules held together by weak forces, this mosaicity is more pronounced than in crystals of rigid organic or inorganic molecules, and the reflections from protein crystals suffer a greater mosaic spread than do those from more ordered crystals.

C. Multiple crystalline forms

In efforts to obtain crystals, or to find optimal conditions for crystal growth, crystallographers sometimes obtain a protein or other macromolecule in more than one crystalline form. Compare, for instance, Figs. 3.1a and 3.1e, which

---

Figure 3.1   Some protein crystals grown by a variety of techniques and using a number of different precipitating agents. They are (a) deer catalase, (b) trigonal form fructose-1,6-diphosphatase from chicken liver, (c) cortisol binding protein from guinea pig sera, (d) concanavalin B from jack beans, (e) beef liver catalase, (f) an unknown protein from pineapples, (g) orthorhombic form of the elongation factor Tu from *Escherichia coli,* (h) hexagonal and cubic crystals of yeast phenylalanine tRNA, (i), monoclinic laths of the gene 5 DNA unwinding protein from bacteriophage fd, (j) chicken muscle glycerol-3-phosphate dehydrogenase, and (k) orthorhombic crystals of canavalin from jack beans. From A. McPherson, in *Methods in Enzymology* **114,** H. W. Wyckoff, C. H. W. Hirs, and S. N. Timasheff, eds., Academic Press, Orlando, Florida, 1985, p. 114. Photo generously provided by the author; photo and caption reprinted with permission.

show crystals of the enzyme catalase from two different species. Although these enzymes are almost identical in molecular structure, they crystallize in different forms. In Fig. 3.1h, you can see that highly purified yeast phenylalanyl tRNA (transfer ribonucleic acid) crystallizes in two different forms. Often, the various crystal forms will differ in quality of diffraction, in ease and reproducibility of growth, and perhaps in other properties. The crystallographer must ultimately choose the best form with which to work. Quality of diffraction is the most important criterion, because it determines the ultimate quality of the crystallographic model. Among forms that diffract equally well, more symmetrical forms are usually preferred because they require less data collection (see Chapter 4).

D. Water content

Early protein crystallographers, proceeding by analogy with studies of other crystalline substances, examined dried protein crystals and obtained no diffraction patterns. Thus X-ray diffraction did not appear to be a promising tool for analyzing proteins. In 1934, J. D. Bernal and Dorothy Crowfoot (later Hodgkin) measured diffraction from pepsin crystals still in the mother liquor. Bernal and Crowfoot recorded sharp diffraction patterns, with reflections out to distances in reciprocal space that correspond in real space to the distances between atoms. The announcement of their success was, in effect, a birth announcement for protein crystallography.

Careful analysis of electron-density maps usually reveals many ordered water molecules on the surface of crystalline proteins (Plate 4). Additional disordered water is presumed to occupy regions of low density between the ordered particles. The quantity of water varies among proteins and even among different crystal forms of the same protein. The number of detectable ordered water molecules averages about one per amino-acid residue in the protein. Both the ordered and disordered water are essential to crystal integrity, and drying destroys the crystal structure. For this reason, protein crystals are subjected to X-ray analysis in a very humid atmosphere or in a solution that will not dissolve them, such as the mother liquor.

NMR analysis of protein structure suggests that the ordered water molecules seen by X-ray diffraction on protein surfaces have very short residence times in solution. Thus most of these molecules may be of little importance to an understanding of protein function. However, ordered water is of great importance to the crystallographer. As the structure determination progresses, ordered water becomes visible in the electron-density map. For example, in Plate 2, water molecules are implied by small regions of disconnected density.

Positions of these molecules are indicated by red crosses. Assignment of water molecules to these isolated areas of electron density improves the overall accuracy of the model, and for reasons I will discuss in Chapter 7, improvements in accuracy in one area of the model gives accompanying improvements in other regions.

II. Evidence that solution and crystal structures are similar

Knowing that crystallographers study proteins in the crystalline state, you may be wondering if these molecules are altered when they crystallize, and whether the structure revealed by X rays is pertinent to the molecule's action in solution. Crystallographers worry about this problem also, and with a few proteins, it has been found that crystal structures are in conflict with chemical or spectroscopic evidence about the protein in solution. These cases are rare, however, and the large majority of crystal structures appear to be identical to the solution structure. Because of the slight possibility that crystallization will alter molecular structure, an essential part of any structure determination project is an effort to show that the crystallized protein is not significantly altered.

A. Proteins retain their function in the crystal

Probably the most convincing evidence that crystalline structures can safely be used to draw conclusions about molecular function is the observation that many macromolecules are still functional in the crystalline state. For example, substrates added to suspensions of crystalline enzymes are converted to product, albeit at reduced rates, suggesting that the enzyme's catalytic and binding sites are intact. The lower rates of catalysis can be accounted for by the reduced accessibility of active sites within the crystal, in comparison to solution.

In a dramatic demonstration of the persistence of protein function in the crystalline state, crystals of deoxyhemoglobin shatter in the presence of oxygen. Hemoglobin molecules are known to undergo a substantial conformational change when they bind oxygen. The conformation of oxyhemoglobin is apparently incompatible with the constraints on deoxyhemoglobin in crystalline form, and so oxygenation disrupts the crystal.

It makes sense, therefore, after obtaining crystals of a protein and before embarking on the strenuous process of obtaining a structure, to determine whether the protein retains its function in the crystalline state. If the crystalline form is functional, the crystallographer can be confident that the model will show the molecule in its functional form.

B. X-ray structures are compatible with other structural evidence

Further evidence for the similarity of solution and crystal structures is the compatibility of crystallographic models with the results of chemical studies on proteins. For instance, two reactive groups in a protein might be linked by a cross-linking reagent, demonstrating their nearness. The groups shown to be near each other by such studies are practically always found near each other in the crystallographic model.

In a growing number of cases, both NMR and X-ray methods have been used to determine the structure of the same molecule. Plate 5 shows the alpha-carbon backbones of two models of the protein thioredoxin. The blue model was obtained by X-ray crystallography and the red model by NMR. Clearly the two methods produce similar models. The models are most alike in the pleated-sheet core and the alpha helices. The greatest discrepancies, even though they are not large, lie in the surface loops at the top and bottom of the models. This and other NMR-derived models confirm that protein molecules are very similar in crystals and in solution. In some cases, small differences are seen and can usually be attributed to crystal packing. Often these packing effects are detectable in the crystallographic model itself. For instance, in the crystallographic model of cytoplasmic malate dehydrogenase (PDB file 4mdh), whose functional form is a dimer, an external loop has different conformations in the two molecules of one dimer. On examination of the dimer in the context of neighboring dimers, it can be seen that one molecule of each pair lies very close to a molecule of a neighboring pair. It was thus inferred that the observed difference between the oligomers in a dimer is due to crystal packing, and further, that the unaffected molecule of each pair is probably more like the enzyme in solution.

C. Other evidence

In a few cases, the structure of a protein has been obtained from more than one type of crystal. The resulting models were identical, suggesting that the molecular structure was not altered by crystallization.

Recall that stable protein crystals contain a large amount of both ordered and disordered water molecules. As a result, the proteins in the crystal are still in the aqueous state, subject to the same solvent effects that stabilize the structure in solution. Viewed in this light, it is less surprising that proteins retain their solution structure in the crystal.

III. Growing protein crystals

A. Introduction

Crystals suffer damage in the X-ray beam, primarily due to free radicals generated by X rays. For this reason, a full structure determination project often consumes many crystals. I will now consider the problem of developing a reliable, reproducible source of protein crystals. This entails not only growing good crystals of the pure protein, but also obtaining *derivatives*, or crystals of the protein in complex with various nonprotein components (loosely called ligands). For example, in addition to pursuing the structures of proteins themselves, crystallographers also seek structures of proteins in complexes with ligands such as cofactors, substrate analogs, inhibitors, and allosteric effectors. Structure determination then reveals the details of protein-ligand interactions, giving insight into protein function.

Another vital type of ligand is a heavy-metal atom or ion. Crystals of protein/heavy-metal complexes, often called *heavy-atom derivatives*, are usually needed in order to solve the phase problem mentioned in Chapter 2 (Section VI.F). I will show in Chapter 6 that, for the purpose of obtaining phases, it is crucial that heavy-atom derivatives possess the same unit-cell dimensions and symmetry, and the same protein conformation, as crystals of the pure protein, which in discussions of derivatives are called native crystals. So in most structure projects, the crystallographer must produce both native and derivative crystals under the same or very similar circumstances.

B. Growing crystals: Basic procedure

Crystals of an inorganic substance can often be grown by making a hot, saturated solution of the substance and then slowly cooling it. Polar organic compounds can sometimes be crystallized by similar procedures or by slow precipitation from aqueous solutions by addition of organic solvents. If you work with proteins, just the mention of these conditions probably makes you

Hanging
drop

Cover slip

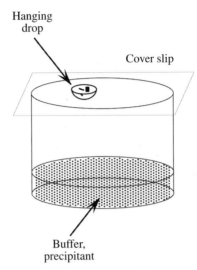

Buffer,
precipitant

Figure 3.2 Growing crystals by the hanging-drop method. The droplet hanging under the cover slip contains buffer, precipitant, protein, and, if all goes well, growing protein crystals.

cringe. Proteins, of course, are usually denatured by heating or exposure to organic solvents, so techniques used for small molecules are not appropriate. In the most common methods of growing protein crystals, purified protein is dissolved in an aqueous buffer containing a precipitant, such as ammonium sulfate or polyethylene glycol, at a concentration just below that necessary to precipitate the protein. Then water is removed by controlled evaporation to produce precipitating conditions, which are maintained until crystal growth ceases.

One widely used technique is vapor diffusion, in which the protein/precipitant solution is allowed to equilibrate in a closed container with a larger aqueous reservoir whose precipitant concentration is optimal for producing crystals. An example of this technique is the hanging-drop method (Fig. 3.2).

Less than 25 μL of the solution of purified protein is mixed with an equal amount of the reservoir solution, giving precipitant concentration about 50% of that required for protein crystallization. This solution is suspended as a droplet underneath a cover glass, which is sealed onto the top of the reservoir grease. Because the precipitant is the major solute present, vapor diffusion in this closed system results in net transfer of water from the protein solution to the reservoir, until the precipitant concentration is the same in both solutions. Because the reservoir is much larger than the protein solution, the final concentration of the precipitant in the protein solution is nearly equal to that in the reservoir. When the system comes to equilibrium, net transfer of water ceases, and the protein solution is maintained at the optimal precipitant concentration. In this way, the precipitant concentration in the protein solution rises to the optimal for crystallization and remains there without overshooting because, at equilibrium, the vapor pressure in the closed system equals the inherent vapor pressure of both protein solution and reservoir.

Frequently the crystallographer obtains many small crystals instead of a few that are large enough for diffraction measurements. If many crystals grow at once, the supply of dissolved protein will be depleted before crystals are large enough to be useful. Small crystals of good quality can be used as seeds to grow larger crystals. The experimental setup is the same as before, except that each hanging droplet is seeded with a few small crystals. Crystals may grow from seeds up to ten times faster than they grow anew, so most of the dissolved protein goes into only a few crystals.

C. Growing derivative crystals

Crystallographers obtain the derivatives needed for phase determination and for studying protein–ligand interactions by two methods—cocrystallizing protein and ligand, and soaking preformed protein crystals in mother-liquor solutions containing ligand.

It is sometimes possible to obtain crystals of protein–ligand complexes by crystallizing protein and ligand together, a process called *cocrystallization*. For example, a number of NAD^+-dependent dehydrogenase enzymes readily crystallize as NAD^+ or NADH complexes from solutions containing these cofactors. Cocrystallization is the only method for producing crystals of proteins in complexes with large ligands, such as nucleic acids or other proteins.

A second means of obtaining crystals of protein–ligand complexes is to soak protein crystals in mother liquor that contains ligand. As mentioned earlier, proteins retain their binding and catalytic functions in the crystalline state, and ligands can diffuse to active sites and binding sites through channels of water in the crystal. Soaking is usually preferred over cocrystallization when the crystallographer plans to compare the structure of a pure protein with that of a protein–ligand complex. Soaking preformed protein crystals with ligands is more likely to produce crystals of the same form and unit-cell dimensions as those of pure protein, so this method is recommended for first attempts to make heavy-atom derivatives.

D. Finding optimal conditions for crystal growth

The two most important keys to success of a crystallographic project are purity and quantity of the macromolecule under study. Impure samples will not make suitable crystals, and even for proteins of the highest purity, repeated trials will be necessary before good crystals result.

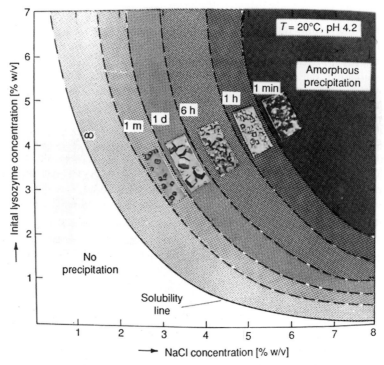

Figure 3.3 Schematic map of crystallization kinetics as a function of lysozyme and NaCl concentration obtained from a matrix of dishes. Inserts show photographs of dishes obtained 1 month after preparation of solutions. From G. Feher and X. Kam, in *Methods in Enzymology* **114,** H. W. Wyckoff, C. H. W. Hirs, and S. N. Timasheff, eds., Academic Press, Orlando, Florida, 1985, p. 90. Photo and caption reprinted with permission.

Many variables influence the formation of macromolecular crystals. These include obvious ones like protein purity, concentrations of protein and precipitant, pH, and temperature, as well as more subtle ones like cleanliness, vibration and sound, convection, source and age of the protein, and the presence of ligands. Clearly, the problem of developing a reliable source of crystals entails controlling and testing a large number of parameters. The difficulty and importance of obtaining good crystals has even prompted the invention of crystallization robots that can be programmed to set up many trials under systematically varied conditions.

The complexity of this problem is illustrated in Fig. 3.3, which shows the effects of varying just two parameters, the concentrations of protein (in this case, the enzyme lysozyme) and precipitant (NaCl). Notice the effect of slight changes in concentration of either protein or precipitant on the rate of crystallization, as well as the size and quality of the resulting crystals.

A sample scheme for finding optimum crystallization conditions is to determine the effect of pH on precipitation with a given precipitant, repeat this determination at various temperatures, and then repeat these experiments with different precipitating agents. Notice in Fig. 3.3, that the region of [protein] versus [precipitant] that gives best crystals is in the shape of an arc. It turns out that if these same data are plotted as [protein] versus ([protein] × [precipitant]), this arc-shaped region becomes a rectangle, which makes it easier to survey the region systematically. For such surveys of crystallization conditions, multiple batches of crystals can be grown conveniently by the hanging-drop method in clear plastic tissue-culture trays of 24 or more wells, each with its own cover glass. This apparatus has the advantage that the growing crystals can be observed through the cover glasses with a dissecting microscope. Then, once the ideal conditions are found, many small batches of crystals can be grown at once, and each batch can be harvested without disturbing the others.

Crystallographers have developed sophisticated schemes for finding and optimizing conditions for crystal growth. One approach, called a response-surface procedure, begins with the establishment of a scoring scheme for results, such as giving higher scores for lower ratios of the shortest to the longest crystal dimension. This gives low scores for needles and higher scores for cubes. Then crystallization trials are carried out, varying several parameters, including pH, temperature, and concentrations of protein, precipitant, and other additives. The results are scored, and the relationships between parameters and scores are analyzed. These relationships are fitted to mathematical functions (like polynomials), which describe a complicated multidimensional surface (one dimension for each variable or for certain revealing combinations of variables) over which the score varies. The crystallographer wants to know the location of the "peaks" on this surface, where scores are highest. Such peaks may lie at sets of crystallization conditions that were not tried in the trials and may suggest new and more effective conditions for obtaining crystals. Finding peaks on such surfaces is just like finding the maximum or minimum in any mathematical function. You take the derivative of the function, set it equal to zero, and solve for the values of the parameters. The sets of values obtained correspond to conditions that lie at the top of mountains on the surface of crystal scores.

An example of this approach is illustrated in Fig. 3.4. The graph in the center is a two-dimensional slice of a four-dimensional surface over which [protein], ([protein] × [precipitant]), pH, and temperature were varied, in attempts to find optimal crystallization conditions for the enzyme tryptophanyl-tRNA synthetase. Note that this surface samples the rectangular region [protein] versus ([protein] × [precipitant]), mentioned earlier. The height of the surface is the score for the crystallization. Surrounding the graph are photos of typical crystals obtained in multiple trials of each set of conditions. None of the trial conditions were near the peak of the surface. The

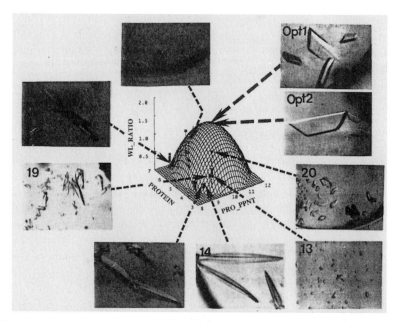

Figure 3.4 Optimization of conditions for crystallization of tryptophanyl-tRNA synthetase. Photo insets show crystals obtained from various conditions represented by points on the surface. Coordinates of the surface are protein concentration (PRO-TEIN), product of protein concentration and precipitant concentration (PRO_PPNT), and the shape of the crystal as reflected by the ratio of its two smallest dimensions, width and length (WL_RATIO). From C. W. Carter, in *Methods in Enzymology* **276,** C. W. Carter and R. M. Sweet, eds., Academic Press, New York, 1997, p. 75. Reprinted with permission.

photos labeled Opt1 and Opt2 are of crystals obtained from conditions defined by the surface peak. In this instance, the response-surface approach predicted conditions that produced better crystals than any from the trials that pointed to these conditions.

When varying the more conventional parameters fails to produce good crystals, the crystallographer may take more drastic measures. Sometimes limited digestion of the protein by a proteolytic enzyme removes a disordered surface loop, resulting in a more rigid, hydrophilic, or compact molecule that forms better crystals. A related measure is adding a ligand, such as a cofactor, that is known to bind tightly to the protein. The protein/cofactor complex may be more likely to crystallize than the free protein, either because the complex is more rigid than the free protein or because the cofactor induces a conformational change that makes the protein more amenable to crystallizing.

Many membrane-associated proteins will not dissolve in aqueous buffers and tend to form amorphous precipitates instead of crystals. The intractability of such proteins often results from hydrophobic domains or surface regions that are normally associated with the interior of membranes. Such proteins have sometimes been crystallized in the presence of detergents, which coat the hydrophobic portion, decorating it with the ionic groups of detergent, and thus rendering it more soluble in water. Also, limited proteolysis of membrane-associated proteins might remove an exposed hydrophobic portion, leaving a crystallizable fragment that is more like a typical water-soluble protein.

The effects of these modifications, as well as the potential crystallizability of a newly purified protein, can be tentatively assessed before crystallization trials begin, through analysis of laser light scattering by solutions of the macromolecule. Simple, rapid light-scattering experiments (see Chapter 9, Section III) can reveal much about the nature of the substance in solutions of varied composition, pH, and temperature, including estimates of average molecular mass of the particles, radius of gyration (dependent on shape of particles), rates of diffusion through the solution, and range and distribution of particle sizes (degree of *polydispersity*). Some of the measured properties correlate well with crystallizability. In particular, *monodisperse* preparations—those containing particles of uniform size—are more promising candidates for crystallization than those in which the protein is polydisperse. In many cases, polydispersity arises from nonspecific interactions among the particles, which at higher concentrations is likely to result in random aggregation rather than orderly crystallization.

When these drastic measures are required to yield good crystals, the crystallographer is faced with the question of whether the resulting fragment is worthy of the arduous effort to determine its structure. This question is similar to the basic issue of whether a protein has the same structure in crystal and in solution, and the question must be answered in the same way. Specifically, it may be possible to demonstrate that the fragment maintains at least part of the biological function of the intact molecule, and further, that this function is retained after crystallization.

IV. Judging crystal quality

The acid test of a crystal's suitability for structure determination is, of course, its capacity to give sharp diffraction patterns with clear reflections at large angles from the X-ray beam. A brief inspection of crystals under a low-power light microscope can also provide some insight into quality and can help the crystallographer pick out the most promising crystals.

Desirable visible characteristics of crystals include optical clarity, smooth faces, and sharp edges. Broken or twinned crystals sometimes exhibit dark cleavage planes within an otherwise clear interior. Depending on the lattice type (Chapter 4) and the direction of viewing relative to unit-cell axes, some crystals strongly rotate plane-polarized light. This property is easily observed by examining the crystal between two polarizers, one fixed and one rotatable, under a microscope. Upon rotation of the movable polarizer, a good-quality crystal will usually brighten and darken sharply.

Another useful physical property of the crystal is its density, which can be used to determine several useful microscopic properties, including the protein molecular weight, the protein/water ratio in the crystal, and the number of protein molecules in each asymmetric unit (defined later). Molecular weights from crystal density are more accurate than those from electrophoresis or most other methods (except mass spectrometry) and are not affected by dissociation or aggregation of protein molecules. The protein/water ratio is used to clarify electron-density maps prior to interpretation (Chapter 7). If the unit cell is symmetric (Chapter 4), it can be subdivided into two or more equivalent parts called *asymmetric units* (the simplest unit cell contains, or in fact is, one asymmetric unit). For interpreting electron-density maps, it is helpful to know the number of protein molecules per asymmetric unit.

Crystal density is measured in a graduated cylinder by suspending the crystal in a density gradient made by mixing water-saturated organic solvents such as xylene and carbon tetrachloride. The crystal will settle through the liquid until its density matches that of the liquid mixture and then remain suspended there. Drops of salt solutions of known density are used to calibrate the gradient.

The product of the crystal density and the unit-cell volume (determined from crystallographic analysis, Chapter 4) gives the total mass within the unit cell. This quantity, expressed in daltons, is the sum of all atomic masses in one unit cell. If the protein molecular mass and the number of protein molecules per unit cell are known, then the remainder of the cell can be assumed to be water, thus establishing the protein/water ratio.

It can be shown that the molecular weight of protein in each asymmetric unit is given by

$$M_P = \frac{N \cdot V (D_C - D_W)}{n(1 - v_P D_W)},$$ (3.1)

in which D_C and D_W are densities of crystal and water, N is Avogadro's number, V is the volume of the unit cell, v_P is the partial specific volume of the protein, and n is the number of protein molecules of molecular mass M_P in each unit cell. The partial specific volume of the protein can be determined from its amino-acid composition (percent of each amino acid) by simply averaging the partial specific volumes of the component amino acids (obtained from tables). Thus if the protein molecular weight is known, n can be computed. Because n is an integer, it can be determined from even a rough molecular weight, taking the integer nearest the computed result. Then substitution of the correct integral value of n into Eq. (3.1) gives a precise value of M_P.

Once the crystallographer has a reliable source of suitable crystals, data collection can begin.

V. Mounting crystals for data collection

The classical method of mounting crystals is to transfer them into a fine glass capillary along with a droplet of the mother liquor. The capillary is then sealed at both ends and mounted onto a goniometer head (see Fig. 4.20, and Chapter 4, Section III.D), a device that allows control of the crystal's orientation in the X-ray beam. The droplet of mother liquor keeps the crystal hydrated.

For many years, crystallographers have been aware of the advantages of collecting X-ray data on crystals at very low temperatures, such as that of liquid nitrogen (boiling point $-196°$ C). In theory, lowering the temperature should increase molecular order in the crystal and improve diffraction. In practice, however, early attempts to freeze crystals resulted in damage due to formation of ice crystals. In recent years, crystallographers have developed techniques for flash freezing crystals in the presence of agents like glycerol, which prevent ice from forming. Crystallography at low temperatures is called *cryocrystallography* and the ice-preventing agents are called *cryoprotectants*. Other cryoprotectants include xylitol or sugars such as glucose. Some precipitants, for example, polyethylene glycol, act as cryprotectants, and often it is only necessary to increase their concentration in order to achieve protection from ice formation.

Preparation of crystals for cryocrystallography typically entails placing them in a cryoprotected mother liquor for 5–15 seconds to wash off the old mother liquor. Then they are immediately frozen. The best final concentration of protectant is determined by trial and error, but often a wide range of concentrations will work. After transfer into protectant, the crystal is picked up in a small (< 1 mm) circular loop of nylon fiber or glass wool, where it remains suspended in a thin film of solvent, sort of like the soap film in a plastic loop for blowing soap bubbles. The loop is then dipped into liquid nitrogen. If flash-freezing is successful, the liquid film in the loop freezes into a glass and remains clear. For data collection, the loop is mounted onto the goniometer, where it is held in a stream of cold nitrogen gas coming from a reservoir of liquid nitrogen. A temperature of $-100°$ C can be maintained in this manner.

In addition to better diffraction, other benefits of cryocrystallography include reduction of radiation damage to the crystal and hence the possibility of collecting more data from a single crystal; reduction of X-ray scattering from water (resulting in cleaner backgrounds in diffraction patterns) because the amount of water surrounding the crystal is far less than that in a droplet of mother liquor in a capillary; and the possibility of safe storage, transport, and reuse of crystals. Crystallographers can take or ship loop-mounted flash-frozen crystals to sites of data collection, minimizing handling of crystals at the collection site. With all these potential benefits, it is not surprising that cryocrystallography has become common practice.

4 Collecting Diffraction Data

I. Introduction

In this chapter, I will discuss the geometric principles of diffraction, revealing, in both the real space of the crystal's interior and in reciprocal space, the conditions that produce reflections. I will show how these conditions allow the crystallographer to determine the dimensions of the unit cell and the symmetry of its contents and how these factors determine the strategy of data collection. Finally, I will look at the devices used to produce and detect X rays and to measure precisely the intensities and positions of reflections.

II. Geometric principles of diffraction

W. L. Bragg showed that the angles at which diffracted beams emerge from a crystal can be computed by treating diffraction as if it were reflection from sets of equivalent, parallel planes of atoms in a crystal. (This is why each spot in the diffraction pattern is called a *reflection*.) I will first describe how crystallographers denote the planes that contribute to the diffraction pattern.

A. The generalized unit cell

The dimensions of a unit cell are designated by six numbers: the lengths of three unique edges **a**, **b**, and **c**; and three unique angles α, β, and γ (Fig. 4.1). Notice the use of bold type in naming the unit cell edges or the axes that correspond to them. I will use bold letters (**a**, **b**, **c**) to signify the edges or axes themselves, and letters in italics (*a, b, c*) to specify their length. Thus *a* is the length of unit cell edge **a**, and so forth.

A cell in which $a \neq b \neq c$ and $\alpha \neq \beta \neq \gamma$, as in Fig. 4.1, is called *triclinic*. If $a \neq b \neq c$, $\alpha = \gamma = 90°$, and $\beta > 90°$, the cell is *monoclinic*. If $a = b$, $\alpha = \beta = 90°$, and $\gamma = 120°$, the cell is *hexagonal*. For cells in which all three cell angles are 90°, if $a = b = c$, the cell is *cubic*; if $a = b \neq c$, the cell is *tetragonal*; and if $a \neq b \neq c$, the cell is *orthorhombic*. The most convenient coordinate systems for crystallography adopt coordinate axes based upon the directions of unit-cell edges. For cells in which at least one cell angle is not 90°, the coordinate axes are not the familiar orthogonal (mutually perpendicular) *x, y,* and *z*. In this book, for clarity, I will consider only unit cells and coordinate systems with orthogonal axes ($\alpha = \beta = \gamma = 90°$), and I will use orthorhombic systems most often, making it easy to distinguish the three cell edges. In such systems, the **a** edges of the cell are parallel to the *x*-axis of an orthogonal coordinate system, edges **b** are parallel to *y*, and edges **c** are parallel to *z*. Bear in mind, however, that the principles discussed here can be generalized to all unit cells.

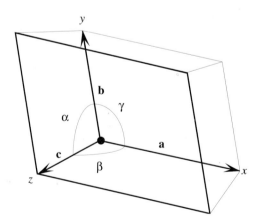

Figure 4.1   General (triclinic) unit cell, with edges **a**, **b**, **c** and angles α, β, γ.

B. Indices of the atomic planes in a crystal

The most readily apparent sets of planes in a crystalline lattice are those determined by the faces of the unit cells. These and all other regularly spaced planes that can be drawn through lattice points can be thought of as sources of diffraction and can be designated by a set of three numbers called *lattice indices*. Three indices *hkl* identify a particular set of equivalent, parallel planes. The index *h* gives the number of planes in the set per unit cell in the *x* direction or, equivalently, the number of parts into which the set of planes cut the **a** edge of each cell. The indices *k* and *l* specify how many such planes exist per unit cell in the *y* and *z* directions. An equivalent way to determine the indices of a set of planes is to start at any lattice point and move out into the unit cell away from the plane cutting that lattice point. If the first plane encountered cuts the **a** edge at some fraction $1/n_a$ of its length, and the same plane cuts the **b** edge at some fraction $1/n_b$ of its length, then the *h* index is n_a and the *k* index is n_b (examples are given later). Indices are written in parentheses when referring to the set of planes; hence, the planes having indices *hkl* are the (*hkl*) planes.

In Fig. 4.2, each face of an orthorhombic unit cell is labeled with the indices of the set of planes that includes that face. (The crossed arrows lie upon the labeled face.)

The set of planes including and parallel to the **bc** face, and hence normal to the *x*-axis, is designated (100) because there is one such plane per lattice point in the *x* direction. In like manner, the planes parallel to and including the **ac** face are called (010) planes (one plane per lattice point along *y*). Finally, the **ab** faces of the cell determine the (001) planes. In the Bragg model of diffraction as reflection from parallel sets of planes, any of these sets of planes can be the source of one diffracted X-ray beam. (Remember that an entire set of parallel planes, not just one plane, acts as a single diffractor and produces one

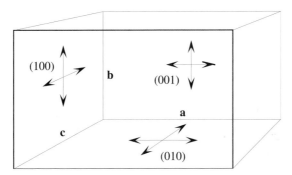

Figure 4.2 Indices of faces in an orthorhombic unit cell.

reflection.) But if these three sets of planes were the only diffractors, the number of diffracted beams would be small, and the information obtainable from diffraction would be very limited.

In Fig. 4.3, an additional set of planes, and thus an additional source of diffraction, is indicated. The lattice (dark lines) is shown in section parallel to the **ab** faces or the *xy* plane. The dashed lines represent the intersection of a set of equivalent, parallel planes that are perpendicular to the *xy* plane of the paper. Note that the planes cut each **a** edge into two parts and each **b** edge into one part, so these planes have indices 210. Because all (210) planes are parallel to the *z* axis (which is perpendicular to the plane of the paper), the *l* index is zero. [Or equivalently, because the planes are infinite in extent, and are coincident with c edges, and thus do not cut edges parallel to the *z* axis, there are zero (210) planes per unit cell in the *z* direction.] As another example, for any plane in the set shown in Fig. 4.4, the first plane encountered from any lattice point cuts that unit cell at *a*/2 and *b*/3, so the indices are 230.

All planes perpendicular to the *xy* plane have indices *hk*0. Planes perpendicular to the *xz* plane have indices *h*0*k,* and so forth. Many additional sets of planes are not perpendicular to *x*, *y*, or *z*. For example, the (234) planes cut the unit cell edges **a** into two parts, **b** into three parts and **c** into four parts. (See Fig. 4.5.)

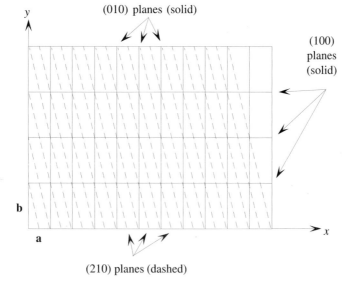

Figure 4.3   (210) planes in a two-dimensional section of lattice.

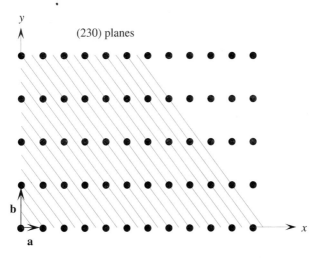

Figure 4.4   (230) planes in a two-dimensional section of lattice.

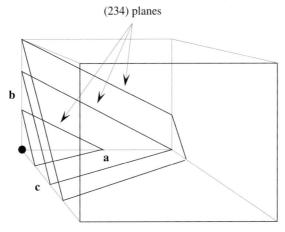

Figure 4.5   The intersection of three (234) planes with a unit cell. Note that the (234) planes cut the unit-cell edges **a** into two parts, **b** into three parts and **c** into four parts.

Finally indices can be negative as well as positive. The (210) planes are the same as (−2 −1 0), whereas the (2 −1 0) or (−2 1 0) planes tilt in the direction opposite to the (210) planes (Fig. 4.6). (The negative signs are often printed on top of the indices, but for clarity I will present them as shown here.)

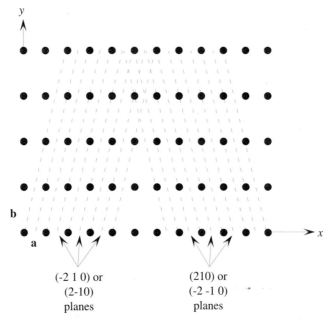

Figure 4.6   The (210) and (−2 −1 0) planes are identical. The (210) planes tilt in the opposite direction from (−1 2 0) and (−2 1 0) planes.

In Bragg's way of looking at diffraction as reflection from sets of planes in the crystal, each set of parallel planes described here (as well as each additional set of planes interleaved between these sets) is treated as an independent diffractor and produces a single reflection. This model is useful for determining the geometry of data collection. Later, when I discuss structure determination, I will consider another model in which each atom or each small volume element of electron density is treated as an independent diffractor, represented by one term in a Fourier series that describes each reflection. Bragg's model tells us where to look for the data. The Fourier series model tells us what the data has to say about molecular structure.

C. Conditions that produce diffraction: Bragg's law

Notice that the different sets of equivalent parallel planes in the preceding figures have different interplanar spacing d. Among sets of planes (*hkl*), interplanar spacing decreases as any index increases. W. L. Bragg showed that a set of parallel planes with index hkl and interplanar spacing d_{hkl} produces a diffracted beam when X rays of wavelength λ impinge upon the planes at an angle θ and are reflected at the same angle, only if θ meets the condition

$$2d_{hkl}\sin\theta = n\lambda, \qquad (4.1)$$

where n is an integer. The geometric construction in Fig. 4.7 demonstrates the conditions necessary for producing a strong diffracted ray. The dots represent two parallel planes of lattice points with interplanar spacing d_{hkl}. Two rays R_1 and R_2 are reflected from them at angle θ.

Lines AC are drawn from the point of reflection A of R_1 perpendicular to the ray R_2. If ray R_2 is reflected at B, then the diagram shows that R_2 travels the same distance as R_1 plus an added distance $2BC$. Because AB in the small triangle ABC is perpendicular to the atomic plane, and AC is perpendicular to the incident ray, the angle CAB equals θ, the angle of incidence. (Two angles are equal if corresponding sides are perpendicular.) Because ABC is a right triangle, the sine of angle θ is BC/AB or BC/d_{hkl}. Thus BC equals $d_{hkl}\sin\theta$, and the additional distance $2BC$ traveled by ray R_2 is $2d_{hkl}\sin\theta$.

If this difference in path length for rays reflected from successive planes is equal to an integral number of wavelengths of the impinging X rays (that is, if $2d_{hkl}\sin\theta = n\lambda$), then the rays reflected from successive planes emerge from the crystal in phase with each other, interfering constructively to produce a strong diffracted beam. For other angles of incidence θ' (where $2d_{hkl}\sin\theta'$

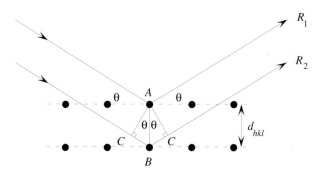

1. $\sin\theta = BC/AB$

2. $BC = AB\sin\theta = d_{hkl}\sin\theta$

Figure 4.7 Conditions that produce strong diffracted rays. If the additional distance traveled by the more deeply penetrating ray R_2 is an integral multiple of λ, then rays R_1 and R_2 interfere constructively.

does not equal an integral multiple of λ), waves emerging from successive planes are out of phase, so they interfere destructively, and no beam emerges at that angle. Think of it this way: If X-rays impinge at an angle θ' that does not satisfy the Bragg conditions, then for every reflecting plane p, there will exist, at some depth in the crystal, another parallel plane p' producing a wave precisely 180° out of phase with that from p, and thus precisely cancelling the wave from p. So all such waves will be cancelled by destructive interference, and no diffracted ray will emerge at the angle θ'. Strong diffracted rays emerge from (hkl) planes of spacing d_{hkl} only at angles θ for which $2d_{hkl} \sin\theta = n\lambda$.

Notice that the angle of diffraction θ is inversely related to the interplanar spacing d_{hkl} ($\sin\theta$ is proportional to $1/d_{hkl}$). This implies that large unit cells, with large spacings, give small angles of diffraction and hence produce many reflections that fall within a convenient angle from the incident beam. On the other hand, small unit cells give large angles of diffraction, producing fewer measurable reflections. In a sense, the number of measurable reflections depends on how much information is present in the unit cell. Large cells contain many atoms and thus more information, and they produce more information in the diffraction pattern. Small unit cells contain fewer atoms, and diffraction from them contains less information.

It is not coincidental that I use the variable names h, k, and l for both the indices of planes in the crystal and the indices of reflections in the diffraction pattern (Chapter 2, Section V). I will show later that in fact the set of planes (hkl) produces the reflection hkl of the diffraction pattern. In the terms used in Chapter 2, each set of parallel planes in the crystal produces one reflection, or one term in the Fourier series that describes the electron density within the unit cell. The intensity of that reflection depends upon the electron distribution and density along the planes that produce the reflection.

D. The reciprocal lattice

Now let us consider the Bragg conditions from another point of view, in reciprocal space. Before looking at diffraction from this vantage point, I will define and tell how to construct a new lattice, the *reciprocal lattice*, in what will at first seem an arbitrary manner. But I will then show that the points in this reciprocal lattice are guides that tell the crystallographer the angles at which strong reflections will occur.

Figure 4.8a shows an **ab** section of lattice with an arbitrary lattice point O chosen as the origin of the reciprocal lattice I am about to define. This point is thus the origin for both the real and reciprocal lattices. Each + in the figure is a real lattice point.

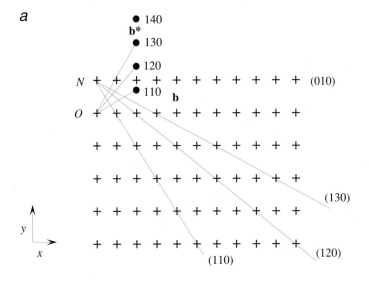

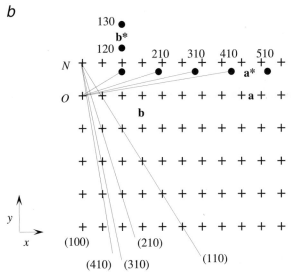

Figure 4.8   (*a*) Construction of reciprocal lattice. Real-lattice points are +s, and reciprocal lattice points are dots. Notice the real cell edges **b** and a reciprocal cell edge **b***. (*b*) Continuation of (*a*). Notice the real cell edges **a** and a reciprocal cell edge **a***.

Through a neighboring lattice point N, draw one plane from each of the sets (110), (120), (130), and so forth. These planes intersect the **ab** section in lines labeled (110), (120), and (130) in Figure 4.8a. From the origin, draw a line normal to the (110) plane. Make the length of this line $1/d_{110}$, the inverse of the interplanar spacing d_{110}. Define the *reciprocal lattice point* 110 as the point at the end of this line (heavy dot). Now repeat the procedure for the (120) plane. drawing a line from O normal to the (120) plane, and of length $1/d_{120}$. Because d_{120} is smaller than d_{110} (recall that d decreases as indices increase), this second line is longer than the first. The end of this line defines a second reciprocal lattice point, with indices 120 (heavy dot). Repeat for the planes (130), (140), and so forth.

Now continue this operation for planes (210), (310), (410), and so on, defining reciprocal lattice points 210, 310, 410, and so on (Fig. 4.8b). Note that the points defined by continuing these operations form a lattice, with the arbitrarily chosen real lattice point as the origin (indices 000). This new lattice is the reciprocal lattice. The planes $hk0$, $h0k$, and $0kl$ correspond, respectively, to the xy, xz, and yz planes. They intersect at the origin and are called the zero-level planes in this lattice. Other planes of reciprocal-lattice points parallel to the zero-level planes are called upper-level planes.

We can also speak of the reciprocal unit cell in such a lattice (Fig. 4.9). If the real unit-cell angles α, β, λ and are 90°, the reciprocal unit cell has axes **a*** lying along real unit cell edge **a**, **b*** lying along **b,** and **c*** along **c**. The lengths of edges **a***, **b***, and **c*** are reciprocals of the lengths of corresponding real cell edges **a**, **b**, and **c** : **a*** = 1/**a**, and so forth. If axial lengths are expressed in angstroms, then reciprocal-lattice spacings are in the unit 1/Å or Å^{-1} (reciprocal angstroms). For real unit cells with nonorthogonal axes, the

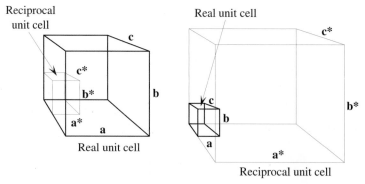

Figure 4.9   Reciprocal unit cells of large and small real cells.

spatial relationships between the real and reciprocal unit-cell edges are more complicated, and I will not make use of them in this book.

Now envision this lattice of imaginary points in the same space occupied by the crystal. For a small real unit cell, interplanar spacings d_{hkl} are small, and hence the lines from the origin to the reciprocal lattice points are long. Therefore, the reciprocal unit cell is large, and lattice points are widely spaced. On the other hand, if the real unit cell is large, the reciprocal unit cell is small, and reciprocal space is densely populated with reciprocal lattice points.

The reciprocal lattice is spatially linked to the crystal because of the way the lattice points are defined, so if we rotate the crystal, the reciprocal lattice rotates with it. So now when you think of a crystal, and imagine the many identical unit cells stretching out in all directions (real space), imagine also a lattice of points in reciprocal space, points whose lattice spacing is inversely proportional to the interplanar spacings within the crystal.

E. Bragg's law in reciprocal space

Now I will look at diffraction from within reciprocal space. I will show that the reciprocal-lattice points give the crystallographer a convenient way to compute the direction of diffracted beams from all sets of parallel planes in the crystalline lattice (real space). This demonstration entails showing how each reciprocal-lattice point must be arranged with respect to the X-ray beam in order to satisfy Bragg's law and produce a reflection from the crystal.

Figure 4.10a shows an $\mathbf{a^*b^*}$ plane of reciprocal lattice. Assume that an X-ray beam (arrow XO) impinges upon the crystal along this plane. Point O is arbitrarily chosen as the origin of the reciprocal lattice. (Remember that O is also a *real-lattice* point in the crystal.) We imagine the X-ray beam passing through O along the line XO (arrow). Draw a circle of radius $1/\lambda$ having its center C on XO and passing through O. This circle represents the wavelength of the X rays in reciprocal space. (If the wavelength is λ in real space, it is $1/\lambda$ in reciprocal space.) Rotating the crystal about O rotates the reciprocal lattice about O, successively bringing reciprocal lattice points like P and P' into contact with the circle. In Fig. 4.10a, P (whose indices are hkl) is in contact with the circle, and the lines OP and BP are drawn. The angle PBO is θ. Because the triangle PBO is inscribed in a semicircle, it is a right triangle and

$$\sin \theta = \frac{OP}{BP} = \frac{OP}{2/\lambda}. \qquad (4.2)$$

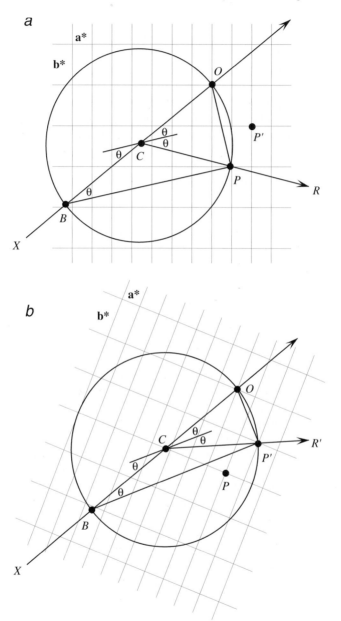

Figure 4.10   Diffraction in reciprocal space. (*a*) Ray *R* emerges from the crystal when reciprocal lattice point *P* intersects the circle. (b) As the crystal rotates, point *P'* intersects the circle, producing ray *R'*.

Rearranging Eq. (4.2) gives

$$2\frac{1}{OP}\sin\theta = \lambda. \tag{4.3}$$

Because P is a reciprocal lattice point, the length of the line OP is $1/d_{hkl}$, where h, k, and l are the indices of the set of planes represented by P. (Recall from the construction of the reciprocal lattice that the length of a line from O to a reciprocal-lattice point hkl is $1/d_{hkl}$). So $1/OP = d_{hkl}$ and

$$2d_{hkl}\sin\theta = \lambda, \tag{4.4}$$

which is Bragg's law with $n = 1$.

In Fig. 4.10b, the crystal, and hence the reciprocal lattice, has been rotated until P', with indices $h'k'l'$, touches the circle. The same construction as in Fig. 4.10a now shows that

$$2d_{h'k'l'}\sin\theta = \lambda. \tag{4.5}$$

We can conclude that whenever the crystal is rotated so that a reciprocal-lattice point comes in contact with this circle of radius $1/\lambda$, Bragg's law is satisfied and a reflection occurs. What direction does the reflected beam take?

Recall (from construction of the reciprocal lattice) that the line defining a reciprocal-lattice point is normal to the set of planes having the same indices as the point. So BP, which is perpendicular to OP, is parallel to the planes that are producing reflection P in Fig. 4.10a. If we draw a line parallel to $B\dot{P}$ and passing through C, the center of the circle, this line (or any other line parallel to it and separated from it by an integral multiple of d_{hkl}) represents a plane in the set that reflects the X-ray beam under these conditions. The beam impinges upon this plane at the angle θ, is reflected at the same angle, and so diverges from the beam at C by the angle 2θ, which takes it precisely through the point P. So CP gives the direction of the reflected ray R in Fig. 4.10a. In Fig. 4.10b, the reflected ray R' follows a different path, the line CP'.

The conclusion that reflection occurs in the direction CP when reciprocal lattice-point P comes in contact with this circle also holds for all points on all circles produce by rotating the circle of radius $1/\lambda$ about the X-ray beam. The figure that results, called the *sphere of reflection*, is shown in Fig. 4.11 intersecting the reciprocal-lattice planes $h0l$ and $h1l$. In the crystal orientation shown, reciprocal-lattice point 012 is in contact with the sphere, so a diffracted ray R is diverging from the source beam in the direction defined by C and point 012. This ray would be detected as the 012 reflection.

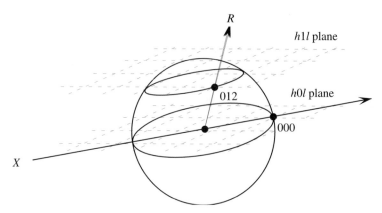

Figure 4.11   Sphere of reflection. When reciprocal lattice point 012 intersects the sphere, ray R emerges from the crystal as reflection 012.

As the crystal is rotated in the X-ray beam, various reciprocal-lattice points come into contact with this sphere, each producing a beam in the direction of a line from the center of the sphere of reflection through the reciprocal-lattice point that is in contact with the sphere. The reflection produced when reciprocal-lattice point P_{hkl} contacts the sphere is called the *hkl* reflection and, according to Bragg's model, is caused by reflection from the set of equivalent, parallel, real-space planes (*hkl*).

This model of diffraction implies that the directions of reflection, as well as the number of reflections, depend only on unit-cell dimensions and not upon the contents of the unit cell. The intensity of reflection *hkl* depends on the values of $\rho(x,y,z)$ on planes (*hkl*). We will see (Chapter 5) that the intensities of the reflections give us the structural information we seek.

F. The number of measurable reflections

If the sphere of reflection has a radius of $1/\lambda$, then any reciprocal-lattice point within a distance $2/\lambda$ of the origin can be rotated into contact with the sphere of reflection (Fig. 4.12).

This distance defines the *limiting sphere*. The number of reciprocal lattice points within the limiting sphere is equal to the the number of reflections that can be produced by rotating the crystal through all possible orientations in the X-ray beam. This demonstrates that the unit-cell dimensions and the wavelength of the X rays determine the number of measurable reflections. Shorter wavelengths make a larger sphere of reflection, bringing more reflections into

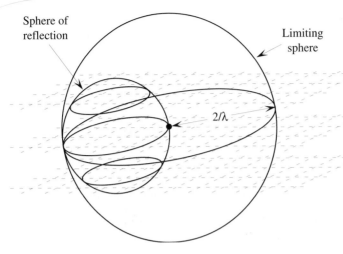

Figure 4.12   Limiting sphere. All reciprocal-lattice points within the limiting sphere of radius $2/\lambda$ can be rotated through the sphere of reflection.

the measurable realm. Larger unit cells mean smaller reciprocal unit cells, which populate the limiting sphere more densely, also increasing the number of measurable reflections.

Because there is one lattice point per reciprocal unit cell (one-eighth of each lattice point lies within each of the eight unit-cell vertices), the number of reflections within the limiting sphere is approximately the number of reciprocal unit cells within this sphere. So the number N of possible reflections equals the volume of the limiting sphere divided by the volume V_{recip} of one reciprocal cell. The volume of a sphere of radius r is $(4\pi/3)r^3$, and r for the limiting sphere is $2/\lambda$, so

$$N = \frac{(4\pi / 3) \cdot (2 / \lambda)^3}{V_{recip}}. \tag{4.6}$$

The volume V of the real unit cell is V_{recip}^{-1}, so

$$N = \frac{33.5 \cdot V}{\lambda^3}. \tag{4.7}$$

Equation (4.7) shows that the number of available reflections depends only upon V and λ. For a modest-size protein unit cell of dimensions 40 x 60 x 80 Å, 1.54-Å radiation can produce 1.76×10^6 reflections, an overwhelming amount of data. Fortunately, because of cell and reciprocal-lattice symmetry, not all of these reflections are unique (Section III.G). Still, getting most of the available information from the diffraction experiment with protein crystals usually requires measuring somewhere between 10^3 and 10^6 reflections.

It can be further shown that the limit of resolution in an image derived from diffraction information is roughly equal to 0.707 times d_{min}, the minimum interplanar spacing that gives a measurable reflection at the wavelength of the x-radiation. For instance, with 1.54-Å radiation, the resolution attainable from all the available data is 0.8 Å, which is more than needed to resolve atoms. A resolution of 1.5 Å, which barely resolves adjacent atoms, can be obtained from about half the available data. Interpretable electron-density maps can usually be obtained with data only out to 2.5 or 3 Å. The number of reflections out to 2.5 Å is roughly the volume of a limiting reciprocal sphere of radius 1/(2.5 Å) multiplied by the volume of the real unit cell. For the unit cell in the preceeding example, this gives about 50,000 reflections. (For a sample calculation, see Chapter 8.)

G. Unit-cell dimensions

Because reciprocal-lattice spacings determine the angles of reflection, the spacings of reflections on the film are related to reciprocal-lattice spacings. (The exact relationship depends on the geometry of recording the reflections, as discussed later.) Reciprocal-lattice spacings, in turn, are simply the inverse of real-lattice spacings. So the distances between reflections on the film and the dimensions of the unit cell are closely connected, making it possible to measure unit-cell dimensions from film spacings. I will discuss the exact geometric relationship in Section III.F, in the context of data-collection devices, whose geometry determines the method of computing unit-cell size.

H. Unit-cell symmetry

If the unit-cell contents are symmetric, then the reciprocal lattice is also symmetric and certain sets of reflections are equivalent. In theory, only one member of each set of equivalent reflections need be measured, so awareness of unit-cell symmetry can greatly reduce the magnitude of data collection. In practice, modest redundancy of measurements improves accuracy, so when more than one equivalent reflection is observed (measured), or when the same reflection is observed more than once, the average of these multiple observations is considered more accurate than any single observation.

In this section, I will discuss some of the simplest aspects of unit-cell symmetry. Crystallography in practice requires detailed understanding of these matters, but users of crystallographic models need only understand their general importance. As we will see later (Section III.G of this chapter and Chapter 5), the crystallographer can determine the unit-cell symmetry from a limited amount of X-ray data and thus can devise a strategy for data collection that will minimize repeated observation of equivalent reflections.

The symmetry of a unit cell is described by its space group, which is represented by a cryptic symbol (like $P2_12_12_1$), in which a capital letter indicates the lattice type and the other symbols represent symmetry operations that can be carried out on the unit cell without changing its appearance. Mathematicians in the late 1800s showed that there are exactly 230 possible space groups.

The unit cells of a few lattice types are shown in Fig. 4.13.

P designates a primitive lattice, containing one lattice point at each corner or vertex of the cell. Because each lattice point is shared among eight neighboring unit cells, a primitive lattice contains eight times one-eighth or one lattice point per unit cell. Symbol I designates a body-centered or *internal* lattice, with an additional lattice point in the center of the cell, and thus two lattice points per unit cell. Symbol F designates a face-centered lattice, with additional lattice points (beyond the primitive ones) on the center of each face.

An example of a symmetry operation is rotation of an object about an axis. To illustrate with a familiar object, if a rectangular table is rotated 180° about an axis perpendicular to and centered on the tabletop (Fig. 4.14), the table looks just the same as it did before rotation (ignoring imperfections such as coffee stains). We say that the table possesses a twofold rotation axis because, in rotating the table one full circle about this axis, we find two positions that are equivalent: 0° and 180°. The axis itself is an example of a symmetry element.

Protein molecules are inherently asymmetric, being composed of chiral amino-acid residues coiled into larger chiral structures such as right-handed helices or twisted beta sheets. If only one protein molecule occupies a unit cell, then the cell itself is chiral, and there are no symmetry elements. This situation is rare; in most cases, the unit cell contains several identical molecules

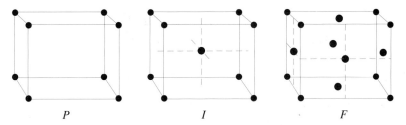

Figure 4.13 P, I, and F unit cells.

or oligomeric complexes in an arrangement that produces symmetry elements. In the unit cell, the largest aggregate of molecules that possesses no symmetry elements, but can be juxtaposed on other identical entities by symmetry operations, is called the *asymmetric unit*. In the simplest case for proteins, the asymmetric unit is a single protein molecule.

The simplest symmetry operations and elements needed to describe unit-cell symmetry are translation, rotation (element: rotation axis), and reflection (element: mirror plane). Combinations of these elements produce more complex symmetry elements, including centers of symmetry, screw axes, and glide planes (discussed later). Because proteins are inherently asymmetric, mirror planes and more complex elements involving them are not found in unit cells of proteins. All symmetry elements in protein crystals are translations, rotations, and screw axes, which are rotations and translations combined.

Translation simply means movement by a specified distance. For example, by the definition of *unit cell*, movement of its contents along one of the unit-cell axes by a distance equal to the length of that axis superimposes the atoms of the cell on corresponding atoms in the neighboring cell. This translation by one axial length is called a *unit translation*. Unit cells often exhibit symmetry elements that entail translations by a simple fraction of axial length, such as $a/4$.

In the space-group symbols, rotation axes such as the twofold axis of the table in Fig. 4.14 are represented in general by the symbol n and specifically by a number. For example, 4 means a fourfold rotation axis. If the unit cell possesses this symmetry element, then it has the same appearance after each $90°$ rotation around the axis.

The screw axis results from a combination of rotation and translation. The symbol nm represents an n-fold screw axis with a translation of m/n of the unit translation. For example, Plate 6 shows models of the amino acid alanine

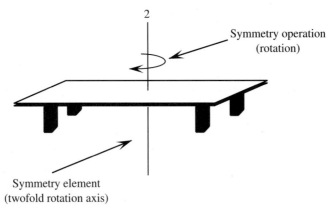

Figure 4.14   Table with a twofold axis of rotation.

on a 3_1 screw axis in a hypothetical unit cell. On the screw axis, each successive molecule is rotated by $120°$ $(360°/3)$ with respect to the previous one, and translated one-third of the axial length.

Plate 7 shows alanine in hypothetical unit cells of two space groups. A triclinic unit cell (Plate 7a) is designated $P1$, being a primitive lattice with only a onefold axis of symmetry (that is, with no symmetry). $P2_1$ (Plate 7b) describes a primitive unit cell possessing a twofold screw axis parallel to $\mathbf{c}$, which points toward you as you view Plate 7. Notice that along any 2_1 screw axis, successive alanines are rotated $180°$ and translated one-half the axis length. A cell in space group $P2_12_12_1$ possesses three perpendicular twofold screw axes.

For the crystallographer, one of the most useful ways to describe unit-cell symmetry is by *equivalent positions*, positions in the unit cell that are superimposed on each other by the symmetry operations. In a $P2_1$ cell with an atom located at (x,y,z), an identical atom can be found at $(-x, -y, {}^1\!/2 + z)$, because the operation of a 2_1 screw axis interchanges these positions. So a $P2_1$ cell has the equivalent positions (x, y, z) and $(-x, -y, {}^1\!/2 + z)$. (The ${}^1\!/2$ means one-half of a unit translation along $\mathbf{c}$, or a distance $c/2$ along the z-axis.)

Lists of equivalent positions for the 230 space groups can be found in *International Tables for X-ray Crystallography*, a reference series that contains an

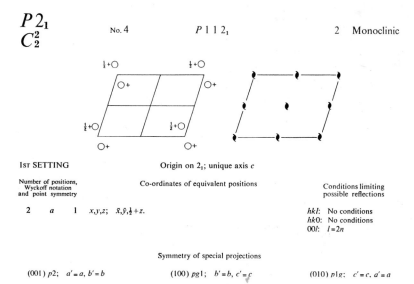

$P2_1$
C_2^2

No. 4 $P\,1\,1\,2_1$ 2   Monoclinic

1ST SETTING   Origin on 2_1; unique axis c

Number of positions, Wyckoff notation and point symmetry			Co-ordinates of equivalent positions	Conditions limiting possible reflections
2	a	1	$x,y,z;\quad \bar{x},\bar{y},\tfrac{1}{2}+z.$	hkl: No conditions $hk0$: No conditions $00l$: $l=2n$

Symmetry of special projections

(001) $p2$; $a'=a, b'=b$ (100) $pg1$; $b'=b, c'=c$ (010) $p1g$; $c'=c, a'=a$

Figure 4.15 $P2_1$ entry in *International Tables for X-ray Crystallography*, Vol. 1, N. F. M. Henry and K. Lonsdale, eds., Reidel NE/Kluwer Academic Publishers, Norwell, Masachusetts, 1969, p. 105. Reprinted with kind permission from Kluwer Academic Publishers and the International Union of Crystallography.

enormous amount of practical information that crystallographers need in their daily work. So the easiest way to see how asymmetric units are arranged in a cell of complex symmetry is to look up the space group in *International Tables*. Each entry contains a list of equivalent positions for that space group, and a diagram of the unit cell. The entry for space group $P2_1$ is shown in Fig. 4.15.

Certain symmetry elements in the unit cell announce themselves in the diffraction pattern by causing specific reflections to be missing (intensity of zero). For example, a twofold screw axis (2_1) along the **c** edge causes all 00*l* reflections having odd values of *l* to be missing. (Notice in Fig. 4.15 that "Conditions limiting possible reflections" in a $P2_1$ cell includes the condition that $l = 2n$, meaning that only the even-numbered reflections are present along the *l*-axis.) As another example, body-centered (*I*) lattices show missing reflections for all values of *hkl* where the sum of *h*, *k*, and *l* is odd. These patterns of missing reflections are called *systematic absences*, and they allow the crystallographer to determine the space group by looking at a few crucial planes of reflections. I will show later in this chapter how symmetry guides the strategy of data collection. In Chapter 5, I will show why symmetry causes systematic absences.

III. Collecting X-ray diffraction data

A. Introduction

Simply stated, the goal of data collection is to determine the indices and record the intensities of as many reflections as possible, as rapidly and efficiently as possible. One cause for urgency is that crystals, especially those of macromolecules, deteriorate in the beam because X rays generate heat and reactive free radicals in the crystal. Thus the crystallographer would like to capture as many reflections as possible during every moment of irradiation. Often the diffracting power of the crystal limits the number of available reflections. Protein crystals that produce measurable reflections from interplanar spacings down to about 3 Å or less are usually suitable for structure determination.

In the following sections, I will discuss briefly a few of the major instruments employed in data collection. These include the X-ray sources, which produce an intense, narrow beam of radiation; detectors, which allow quantitative measurement of reflection intensities; and cameras or diffractometers, which control the orientation of the crystal in the X-ray beam and thus direct reflections having known indices to detectors.

B. X-ray sources

X rays are electromagnetic radiation of wavelengths 0.1–100 Å. X rays in the useful range for crystallography can be produced by bombarding a metal target (most commonly copper or molybdenum) with electrons produced by a heated filament and accelerated by an electric field. A high-energy electron collides with and displaces an electron from a low-lying orbital in a target metal atom. Then an electron from a higher orbital drops into the resulting vacancy, emitting its excess energy as an X-ray photon.

The element in the target exhibits narrow *characteristic lines* (specific wavelengths) of emission resulting from the characteristic energy-level spacing of that element. The wavelengths of emission lines are longer for elements of lower atomic number Z. For instance, electrons dropping from the L shell of copper (Z = 29) to replace displaced K electrons (L $\rightarrow$ K or K_α transition) emit X rays of λ = 1.54 Å. The M $\rightarrow$ K transition produces a nearby emission band (K_β) at 1.39 Å (Fig. 4.16a, solid curve). For molybdenum (Z = 42), $\lambda(K_\alpha)$ = 0.71 Å and $\lambda(K_\beta)$ = 0.63 Å.

A monochromatic (single-wavelength) source of X rays is desirable for crystallography because the diameter of the sphere of reflection is $1/\lambda$, and a source producing two distinct wavelengths of radiation gives two spheres of

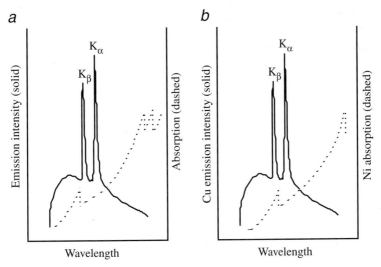

Figure 4.16 (*a*) Emission (solid) and absorption (dashed) spectra of copper. (*b*) Emission spectrum of copper (solid) and absorption spectrum of nickel (dashed). Notice that Ni absorbs K_β more strongly than K_α.

reflection and two interspersed sets of reflections, making indexing difficult or impossible because of overlapping reflections. Elements like copper and molybdenum make good X-ray sources if the weaker K_β radiation can be removed.

At wavelengths away from the characteristic emission lines, each element absorbs X rays. The magnitude of absorption increases with increasing X-ray wavelength and then drops sharply just at the wavelength of K_β. The dashed curve in Fig. 4.16*a* shows the absorption spectrum for copper. The wavelength of this absorption edge, or sharp drop in absorption, like that of characteristic emission lines, increases as Z decreases such that the absorption edge for element Z - 1 lies slightly above the K_β emission line of element Z. This makes element Z - 1 an effective K_β filter for element Z, leaving almost pure monochromatic K_α radiation. For example, a nickel filter 0.015 mm in thickness reduces Cu–K_β radiation to about 0.01 times the intensity of Cu–K_α. Figure 4.16*b* shows the copper emission spectrum (solid) and the nickel absorption spectrum (dashed). Notice that Ni absorbs strongly at the wavelength of Cu–K_β radiation, but transmits Cu–K_α.

There are three common X-ray sources, *X-ray tubes* (actually a cathode ray tube sort of like a television tube), *rotating anode tubes*, and *particle storage rings*, which produce synchrotron radiation in the X-ray region. In the X-ray tube, electrons from a hot filament (cathode) are accelerated by electrically charged plates and collide with a water-cooled anode made of the target metal (Fig. 4.17*a*). X rays are produced at low angles from the anode, and emerge from the tube through windows of beryllium.

Output from X-ray tubes is limited by the amount of heat that can be dissipated from the anode by circulating water. Higher X-ray output can be obtained from rotating anode tubes, in which the target is a rapidly rotating metal disk (Fig. 4.17*b*). This arrangement improves heat dissipation by spreading the electron bombardment over a much larger piece of metal. Rotating anode sources are more than ten times as powerful as tubes with fixed anodes.

Particle storage rings, which are associated with the particle accelerators used by physicists to study subatomic particles, are the most powerful X-ray sources. In these giant rings, electrons or positrons circulate at velocities near the speed of light, driven by energy from radio-frequency transmitters and maintained in circular motion by powerful magnets. A charged body like an electron emits energy (synchrotron radiation) when forced into curved motion, and in accelerators, the energy is emitted as X rays. Accessory devices called *wigglers* cause additional bending of the beam, thus increasing the intensity of radiation. Systems of focusing mirrors and monochromators tangential to the storage ring provide powerful monochromatic X rays at selectable wavelengths.

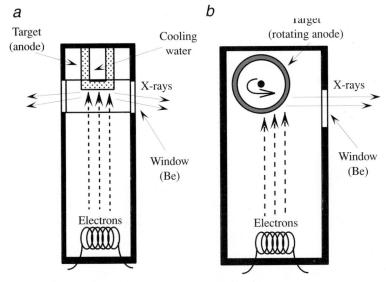

Figure 4.17 (*a*) X-ray tube. (*b*) Rotating anode tube.

A photo and diagram of an accelerator, the Cornell Electron Storage Ring in Ithaca, New York, is shown in Fig. 4.18. The accelerator ring (white in photo and black ring in surrounding diagram) lies buried beneath a soccer field (cut away). The two large buildings lie on opposite sides of the field. Within the larger building is the Cornell High-Energy Synchrotron Source (CHESS), where synchrotron X rays are provided at several workstations. Crystallographers can apply to CHESS and other synchrotron sources for grants of time for data collection. For a detailed virtual tour of CHESS, see the CMCC Home Page on the World Wide Web.

Although synchrotron sources are available only at storage rings and require the crystallographer to collect data away from the usual site of work, there are many advantages that compensate for the inconvenience. X-ray data that requires several hours of exposure to a rotating anode source can often be obtained in seconds or minutes at a synchrotron source like CHESS. In two or three days at a synchrotron source, a crystallographer can collect data that might take months to acquire with conventional sources. Another advantage, as we will see in Chapter 6, is that X rays of selectable wavelength can be helpful in solving the phase problem.

Whatever the source of X rays, the beam is directed through a *collimator*, a narrow metal tube that selects and reflects the X rays into parallel paths, producing a narrow beam. After collimation, beam diameter can be further

Figure 4.18 Cornell Electron Storage Ring. Photo and diagram reprinted with permission of Floyd R. Newman Laboratory of Nuclear Studies, Cornell University. For a virtual tour of CHESS, see the CMCC Home Page on the World Wide Web.

reduced with systems of metal plates called *focusing mirrors*. In the ideal arrangement of source, collimators, and crystal, all points on the crystal can "see" through the collimator and mirrors to all line-of-sight points on the X-ray source.

X-ray sources pose some dangers that the crystallographer must consider in daily work. X-ray tubes require high-voltage power supplies containing large condensers that can produce a dangerous shock even after equipment is shut off. The X rays themselves are relatively nonpenetrating, but can cause serious damage to surface tissues. Even brief exposure to weak X rays can damage eyes, so protective goggles are standard attire in the vicinity of X-ray sources. The direct beam is especially powerful, and sources are electronically interlocked so that the beam entrance shutter cannot be opened while the user is working with the equipment. The beam intensity is always reduced to a minimum during alignment of collimating mirrors or cameras. During data collection, the direct beam is blocked just beyond the crystal by a piece of metal called a *beam stop*, which also has the beneficial effect of preventing excessive radiation from reaching the center of the detector, thus obscuring low-angle reflections. In addition, the entire source, camera, and detector are usually surrounded by Plexiglas to block scattered radiation from the beam stop or collimators but to allow observation of the equipment. As a check on the efficacy of measures to prevent X-ray exposure, the prudent crystallographer wears a dosage-measuring ring or badge during all work with X-ray equipment. These devices are periodically sent to radiation-safety labs for measurement of the X-ray dose received by the worker.

C. Detectors

Reflection intensities can be measured by scintillation counters, which in essence count the X-ray photons and thus give quite accurate intensities over a wide range. Scintillation counters contain a material that produces a flash of light (a scintillation) when it absorbs an X ray photon. A photocell counts the flashes. With simple scintillation counters, each reflection must be measured separately, an arrangement that is convenient only in diffractometry (next section).

The simplest X-ray detector, and for years the workhorse of detectors, is X-ray-sensitive film, but film has been almost completely replaced by image plates and CCD detectors, which are described in more detail later. Various types of cameras (next section) can direct reflections to detectors in useful arrangements, allowing precise determination of indices and intensities for thousands of reflections from a single image.

Image plate detectors can store diffraction images reversibly and have a very wide *dynamic range*, or capacity to record reflections of widely varying intensity. Image plates are plastic sheets with a coating of small crystals of a phosphor, such as $BaF:Eu^{2+}$. The crystals can be stimulated by X rays into a stable excited state in which Eu^{2+} loses an electron to the F^- layer, which contains electron vacancies introduced by the manufacturing process. Further stimulation by visible light causes the electrons to drop back to the Eu layer, producing visible light in proportion to the intensity of the previously absorbed X rays. After X-ray exposure, data are read from the plate by a scanner in which a fine laser beam induces luminiscence from a very small area of the plate, and a photocell records the intensity of emitted light. The intensities are fed to a computer, which can then reconstruct an image of the diffraction pattern. Image plates can be erased by exposure to bright visible light and reused indefinitely.

Area detectors combine the accuracy and wide dynamic range of scintillation counting; the simultaneous measurement of many reflections, as with image plates; and the advantage of direct collection data by computer, without a separate scanning step. One type of area detector is the multiwire or gas-proportional detector. As an example, the Mark I detector at the University of California, San Diego, as diagrammed in Fig. 4.19, consists of two perpendicular sets of parallel wires in a flat box filled with an inert gas. A window of beryllium permits entry of X rays from the front of the detector.

One set of wires, say the x set in Fig. 4.19, is the anode, whereas the perpendicular set is the cathode. The anode set is held at a high positive voltage relative to the cathode set. The wires of each set terminate in a *delay line*, which delays any signal from a wire in its set by a time interval proportional to the distance of the wire from the end of the delay line. This time delay allows determination of which wire produced the signal.

Entering the detector through the beryllium window, an X-ray photon ionizes the gas in a small region (~100 m), producing a few hundred electrons. The electrons drift to the nearest anode wire, and because of the high voltage, each electron triggers an electrical discharge that in turn produces thousands of ion pairs in the gas. The movement of these ions in the electric field of the cathode and anode wires produces a pulse of current in each of the nearest wires. The detection of these pulses at ends of the x and y delay lines allows determination of the reflection position in the detector. A small pulse that appears instantly in the ground connection of the anode delay line serves as a marker against which to time the x and y pulses. While these pulses are moving to the amplifiers, an interval called dead time, the detector is insensitive to ionization events. This sets an upper limit on the detector's counting rate.

Pulses appear in several parallel lines in both the x and y sets, with the strongest pulses in the lines nearest the initial ionization. The known characteristic shape of such a pulse can be fitted to the pulse heights from several neighboring lines, thus locating the ionization event to a higher resolution than the wire spacing in the detector. The output from the area detector is fed to a computer, which indexes the event using the x and y positional information and the crystal orientation at the time of the event. The computer sums events that have the same index and thus produces a file of indexed intensities.

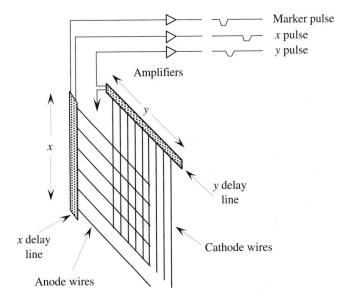

Figure 4.19 Exploded view of detector wires in multiwire area detector.

More recent versions of this design feature an added focusing system, called a radial drift chamber, in front of the detector. Gas ionization occurs within this drift chamber, which is composed of a set of concentric charged rings sandwiched between two lens-shaped metal grids. The cluster of electrons is then focused by the drift chamber, making its direction normal to the surface of the multiwire detector.

There are several advantages to this system. First, the signal shape is the same for X rays detected over the entire detector surface. Without the drift chamber, X rays detected far from the center of the multiwire detector are more spread out because they enter the detector at an angle, making their position less certain. Second, as the electron bunch drifts, it spreads out, ultimately involving more wires in the detection event. Paradoxically, this spreading results in a better sensing of pulse shape, and a better determination of the position of the pulse center. Third, simultaneous X-ray photons are usually distinguishable because the drift time for the two bunches of electrons are likely to differ more than the characteristic dead time of the detector.

The latest designs in area detectors employ charge-coupled devices (CCDs) as detectors. In effect, CCDs are photon counters, solid-state devices that accumulate charge in direct proportion to the amount of light that strikes them. They have found use in astronomy as light collectors of great sensitivity. For crystallographic data collection, CCDs are coated with phosphors that emit visible light in response to X rays. A typical CCD is a 2.5-cm square array of 25-μm pixels, each of which accumulates charge during data collection. A tapered bundle of optical fibers is used to increase the effective detecting area of the array. At the end of a collection cycle, the charges are read out by a process in which rows of pixel charge are transferred sequentially into a serial readout row at one edge of the CCD. The charges in the readout row are transferred serially to an amplifier at the end of the row, and then the next row of pixel charges is transferred into the readout row. Because all data are read out at the end of data collection, a CCD has no dead time, and thus no practical limit on its rate of photon counting.

D. Diffractometers and cameras

Between the irradiated crystal and the detector lies a device for precisely orienting the crystal so as to direct specific reflections toward the detector. In this section, I will describe some of these devices, including the diffractometer, which directs single reflections to a scintillation counter or a larger number of reflections to an area detector; and several cameras, which direct large numbers of reflections simultaneously to film or area detectors. In all cases, the task of these devices is to rotate a crystal through a series of known orientations, causing specified reciprocal lattice points to pass through the sphere of reflection and thus produce diffracted X-ray beams.

In all forms of data collection, the crystal is mounted on a *goniometer head*, a device that allows the crystallographer to set the crystal orientation precisely. The goniometer head (Fig. 4.20) consists of a holder for a capillary tube containing the crystal; two arcs (marked by angle scales), which permit rotation of the crystal by 40° in each of two perpendicular planes; and two dovetailed sledges, which permit small translations of the arcs for centering the crystal on the rotation axis of the head.

Protein crystals, either sealed in capillary tubes with mother liquor or flash-frozen in a fiber loop, are mounted on the goniometer head, which is adjusted to center one face of the crystal perpendicular to the X-ray beam and to allow rotation of the crystal while maintaining centering. Flash-frozen crystals are held in a stream of cold nitrogen gas emerging from a reservoir of liquid nitrogen.

The initial crystal orientations or "settings" are determined by physical examination. Well-formed crystals show distinct faces that are parallel to unit-cell edges, and first attempts to obtain a diffraction pattern are made by placing a crystal face perpendicular to the X-ray beam. Preliminary study of diffraction allows determination of unit-cell dimensions and internal symmetry, as described later. Using this information, the crystallographer decides upon the

Figure 4.20   Goniometer head, with capillary tube holder at top. The tool (right) is an Allen wrench for adjusting arcs and sledges. Photo courtesy of Charles Supper Company.

best set of crystal orientations for collecting data and uses the external faces of the crystal as a guide for orienting unit-cell axes.

In the diffractometer, the goniometer head and crystal are mounted in a system of movable circles called a *goniostat*, which allows automated movement of the crystal into almost any orientation with respect to the X-ray beam and the detector (see Figs. 4.21 and 4.22).

The complete diffractometer consists of a fixed X-ray source, the goniostat, and a movable scintillation-counter detector. The system of circles (Fig. 4.21) allows rotation of the goniometer head (angle ϕ), movement of the head around a circle centered on the X-ray beam (angle χ), and rotation of the χ circle around an axis perpendicular to the beam (angle ω). Furthermore, the detector moves on a circle coplanar with the beam. The axis of this circle coincides with the ω-axis. The position of the detector with respect to the beam is denoted by the angle 2θ. With this arrangement, the crystal can be moved to

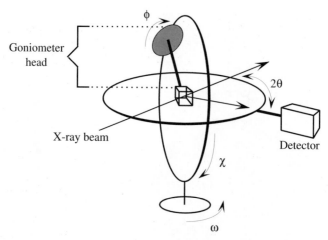

Figure 4.21 System of circles in diffractometry. The crystal in the center is mounted on a goniometer head.

bring any reciprocal lattice point that lies within the limiting sphere into the plane of the detector and into contact with the sphere of reflection, producing diffracted rays in the detector plane. The detector can be moved into proper position to receive, and measure the intensity of, the resulting diffracted beam.

Modern diffractometers are computer-driven and almost completely automated. They can, with some minimal (but important) intervention by the operator, search for reflections and determine unit-cell dimensions and then systematically measure the intensities of all accessible reflections.

This kind of diffractometry gives highly accurate intensity measurement but is slow in comparison with methods that record many reflections at once. In addition, the total irradiation time is long, so crystals may deteriorate and have to be replaced. While one reflection is being recorded, there are usually other unmeasured reflections present, so a considerable amount of diffracted radiation is wasted. Diffractometers can be teamed up with area detectors, as shown in Fig. 4.22, giving substantial increases in the efficiency of data collection.

In this photo, the goniostat (a) and area detector (c) are separated by a drum of helium (b) which transmits X rays with less loss than air. The crystal (d) is barely visible in the tiny glass tube. The two arrows (e) mark the collimator (left) and the beam stop (right), which prevents the direct X-ray beam from reaching the detector. Arrows 1, 2, and 3 on the goniostat indicate the χ, φ, and ω circles.

Of the many types of X-ray cameras, only two are still in even occasional use in protein crystallography—the Buerger or precession camera and the rotation/oscillation camera. The precession camera is used primarily in

Figure 4.22 Diffractometer and area detector. Photo courtesy of Professor Leonard J. Banaszak.

preliminary studies to determine crystal quality, unit-cell dimensions, and symmetry, as well as to assess the quality of derivative crystals. The rotation/oscillation camera is used with film or area detectors to measure large numbers of reflections simultaneously.

The precession camera (Fig. 4.23), although the more complicated in its motion, produces the simplest diffraction pattern. X rays enter through the black tube at left to strike the crystal, mounted in a goniometer head. Beyond the crystal are an annular-screen holder (smaller black square) and a film holder (larger black square). The remaining machinery moves crystal, screen, and film in a precessing motion about the X-ray beam.

Precession photographs reveal the reflections in an undistorted image of the reciprocal lattice. Figure 4.24a shows the geometry that allows precession photography. If the crystal is aligned with one of the real-space axes (a so-called direct axis, as opposed to a reciprocal axis) parallel to the beam, then the zero-level reciprocal-lattice plane along that axis is tangent to the sphere of reflection at the origin. For example, in an orthorhombic system with the **c**-axis (and hence also the **c***-axis) parallel to the beam, the $hk0$ plane is tangent to

Figure 4.23 Precession camera with mounted goniometer head. To see a precession photograph, refer to Fig. 2.6. Photo courtesy of Charles Supper Company.

the sphere at the origin. If the crystal is then inclined slightly, as in Fig. 4.24, this zero-level lattice intersects the sphere of reflection in a circle (small circle in the figure), and any reciprocal-lattice points on this circle produce reflections whose paths lie on a cone radiating from the center of the sphere of reflection through the intersecting circle. If the crystal then precesses about the beam axis at this same angle of inclination, and the film is made to precess about the beam axis in the same manner, the zero-level reflections ($hk0$ in this case) will fall on the film in an undistorted projection of their spatial relationship in the reciprocal lattice. Of course, other nonzero-level points will also intersect the sphere, but their reflections emerge on smaller or larger cones and can be eliminated by interposing an annular screen between crystal and film (Fig. 4.24b). This screen also precesses so that its open annulus always transmits the cone of zero-level reflections. The result is a diffraction pattern recorded as in Chapter 2, Fig. 2.6.

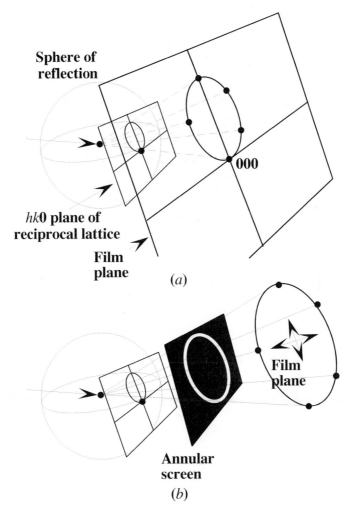

Figure 4.24   (*a*) Geometry of precession photography. Reflections from a single plane of the reciprocal lattice emerge from the crystal on the surface of a cone. (*b*) Annular screen selects one cone of reflections, and thus one plane of reciprocal-lattice points, for photography.

Indexing the resulting photograph is straightforward. In our orthorhombic system, with c^* precessing about the beam, if a^* lies along the x-axis and b^* along y, then the reflection directly to the right of the origin is the 100 reflection, and the one just above the origin is 010.

The crystal is first aligned visually. If the crystal is not perfectly aligned with direct axis precessing about the beam, the image of the reciprocal lattice

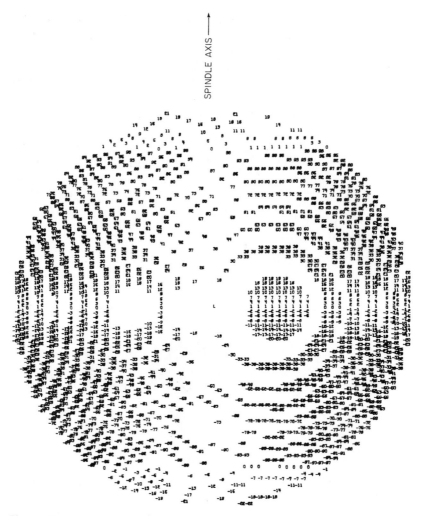

Figure 4.25   Diagram showing expected positions of reflections in an oscillation photograph. Diagram courtesy of Professor Michael Rossmann.

will be distorted. Precise measurements of the distortion provide information that allows the alignment to be corrected. The resulting undistorted image of the reciprocal lattice allows simple measurement of unit-cell dimensions, as discussed in the next section.

Precession cameras are complex but give the diffraction pattern in its simplest, most understandable form. Rotation/oscillation cameras are far simpler, merely providing means to rotate the crystal about an axis perpendicular to the beam, as well as to oscillate it back and forth by a few degrees about the same axis. This movement casts large numbers of reflections in a complex pattern onto the film or area detector (Fig. 4.25).

To produce this figure, a computer program calculated the indices of reflections expected during an oscillation photograph in which the crystal is oscillated about its **c**-axis. At the expected position of each reflection, the program plotted the indices of that reflection. Only the l index of each reflection is shown here, revealing that reflections from many levels of reciprocal space are recorded at once. Although oscillation photographs are very complex, once the unit-cell dimensions are known, all reflections can be indexed.

As the crystal oscillates about a fixed starting position, a limited number of reciprocal-lattice points pass back and forth through the sphere of reflection, and their intensities are recorded. The amount of data from a single oscillation range is limited only by overlap of reflections. The strategy is to take photographs by oscillating the crystal through a small angle about a starting position of rotation, recording all the resulting reflections, and then rotating the crystal to a new starting point such that the new oscillating range overlaps the previous one slightly. From this new position, oscillation produces additional reflections. This process is continued until all reflections have been recorded.

Oscillation photography can be used in tandem with area detectors, but detector resolution limits the size of the unit cell from which data can be collected. Large unit cells, such as those of virus capsids, mean small reciprocal unit cells and large numbers of closely spaced reflections. Image plates are commonly used in such cases because of their greater spatial resolution.

E. Scaling and postrefinement of intensity data

The goal of data collection is a set of consistently measured, indexed intensities for as many of the reflections as possible. After data collection, the raw intensities must be processed to improve their consistency and to maximize the number of measurements that are sufficiently accurate to be used.

A complete set of measured intensities often includes distinct blocks of data obtained from several (or many) crystals and, if data are collected on film, from many films. Because of variability in the diffracting power of crystals, the intensity of the X-ray beam, and the sensitivity of films (if used), the crystallographer cannot assume that the absolute intensities are consistent from one block of data to the next. An obvious way to obtain this consistency is to compare reflections of the same index that were measured from more than one crystal or on more than one film and to rescale the intensities of the two blocks of data so that identical reflections are given identical intensities. This process is called *scaling*. With films, scaling is often preliminary to a more complex process, *postrefinement*, which recovers usable data from reflections that were only partially measured.

Primarily because real crystals are mosaics of submicroscopic crystals (Chapter 3, Section I.B), a reciprocal-lattice point acts as a small three-dimensional entity (sphere or ovoid) rather than as an infinitesimal point. As a reciprocal-lattice point moves through the sphere, diffraction is weak at first, peaks when the center of the point lies precisely on the sphere, and then weakens again before it is extinguished. Accurate measurement of intensity thus entails recording the X-ray output during the entire passage of the point through the sphere. Any range of oscillation will record some reflections only partially, but these may be recorded fully at another rotation angle, allowing partial reflections to be discarded from the data. The problem of partial reflections is serious for large unit cells, where smaller oscillation angles are employed to minimize overlap of reflections. In such cases, if partial reflections are discarded, then a great deal of data is lost.

Data from partial reflections can be interpreted accurately through postrefinement of the intensity data. This process produces an estimate of the partiality of each reflection. Partiality is a fraction p $(0 \geq p \geq 1)$ that can be used as a correction factor to convert the measured intensity of a partial reflection to an estimate of that reflection's full intensity.

Scaling and postrefinement are the final stages in producing a list of internally consistent intensities for most of the available reflections.

F. Determining unit-cell dimensions

The unit-cell dimensions determine the reciprocal-lattice dimensions, which in turn tell us where we must look for the data. Methods like oscillation photography require that we know precisely which reflections will fall completely and partially within a given oscillation angle so that we can collect as many reflections as possible without overlap. So we need the unit-cell dimensions in order to devise a strategy of data collection that will give us as many identifiable (by index), measurable reflections as possible.

Diffractometer software can search for reflections, measure their precise positions, and subsequently compute unit-cell parameters. This search entails complexities we need not encounter here. Instead, I will illustrate the simplest method for determining unit-cell dimensions: measuring reflection spacings from an orthorhombic crystal on a precession photograph.

As discussed earlier, a precession photograph is an undistorted projection of the reciprocal-lattice points onto a flat film. Because reciprocal-lattice spac-

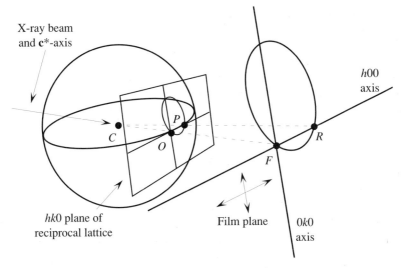

Figure 4.26 Reflection spacings on the film are directly proportional to reciprocal-lattice spacings, and so they are inversely proportional to unit-cell dimensions.

ings are the inverse of real-lattice spacings, the unit-cell dimensions are inversely proportional to the spacing of reflections on a precession photograph. Figure 4.26 shows the geometric relationship between reflection spacings on the film and actual reciprocal-lattice spacings.

The crystal at C is precessing about its $\mathbf{c}^*$-axis, and therefore recording $hk0$ reflections on the film, with the $h00$ axis horizontal and the $0k0$ axis vertical. Point P is the reciprocal-lattice point 100, in contact with the sphere of reflection, and O is the origin. Point F is the origin on the film and R is the recording of reflection 100 on the film. The distance OP is the reciprocal of the distance d_{100}, which is the length of unit-cell edge $\mathbf{a}$. Because CRF and CPO are similar triangles (all corresponding angles equal), and because the radius of the sphere of reflection is $1/\lambda$,

$$\frac{RF}{CF} = \frac{PO}{CO} = \frac{PO}{1/\lambda} = PO \cdot \lambda. \qquad (4.8)$$

Therefore,

$$PO = \frac{RF}{CF \cdot \lambda}. \qquad (4.9)$$

Because $d_{100} = 1/PO$,

$$d_{100} = \frac{CF \cdot \lambda}{RF}. \tag{4.10}$$

In other words, the axial length a (length of unit-cell edge **a**) can be determined by dividing the crystal-to-film distance (CF) by the distance from the film origin to the 100 reflection (RF) and multiplying the quotient by the wavelength of X-rays used in taking the photograph.

In like manner, the vertical reflection spacing along $0k0$ or parallel axes gives $1/d_{010}$, and from it, the length of unit-cell axis **b**. A second precession photograph, taken after rotating this orthorhombic crystal by 90° about its vertical axis, would record the $00l$ axis horizontally, giving $1/d_{001}$ and the length of **c**.

Of course, the distance from the film origin to the 100 reflection on a precession photograph is the same as the distance between any two reflections along this or other horizontal lines, so one photograph allows many measurements to determine accurately the *average* spacing of reciprocal-lattice points along two different axes. From accurate average values, unit-cell-axis lengths can be determined with sufficient accuracy to guide a data-collection strategy.

G. Symmetry and the strategy of collecting data

Strategy of data collection is guided not only by the unit cell's dimensions but also by its internal symmetry. If the cell and its contents are highly symmetric, then certain sets of crystal orientations produce exactly the same reflections, reducing the number of crystal orientations needed in order to obtain all of the distinct or unique reflections.

As mentioned earlier, the unit-cell space group can be determined from systematic absences in the the diffraction pattern. With the space group in hand, the crystallographer can determine the space group of the reciprocal lattice, and thus know which orientations of the crystal will give identical data. All reciprocal lattices possess a symmetry element called a *center of symmetry* or *point of inversion* at the origin. That is, the intensity of each reflection hkl is identical to the intensity of reflection $-h\ -k\ -l$. To see why, recall from our discussion of lattice indices (Section II.B) that the the index of the (230) planes can also be expressed as $(-2\ -3\ 0)$. In fact, the 230 and the $-2\ -3\ 0$ reflections come from opposite sides of the same set of planes, and the reflection intensities are identical. (The equivalence of I_{hkl} and $I_{-h\ -k\ -l}$ is called *Friedel's law*, but there are exceptions. See Chapter 6, Section IV.) This means that half of the reflections in the reciprocal lattice are redundant, and data collection that covers 180° about any reciprocal-lattice axis will capture all unique reflections.

Additional symmetry elements in the reciprocal lattice allow further reduction in the total angle of data collection. It can be shown that the reciprocal lattice possesses the same symmetry elements as the unit cell, plus the additional point of inversion at the origin. The 230 possible space groups reduce to only 11 different groups, called *Laue groups*, when a center of symmetry is added. For each Laue group, and thus for all reciprocal lattices, it is possible to compute the fraction of reflections that are unique. For monoclinic systems, such as P_2, the center of symmetry is the only element added in the reciprocal lattice and the fraction of unique reflections is $1/4$. At the other extreme, for the cubic space group P_{432}, which possesses four-, three-, and twofold rotation axes, only $1/48$ of the reflections are unique. Determination of the crystal symmetry can greatly reduce the number of reflections that must be measured. It also guides the crystallographer in choosing the best axis about which to rotate the crystal during data collection. In practice, crystallographers collect several times as many reflections as the minimum number of unique reflections. They use the redundancy to improve the signal-to-noise ratio by averaging the multiple determinations of equivalent reflections. They also use redundancy to correct for X-ray absorption, which varies with the length of the X-ray path through the crystal.

IV. Summary

The result of X-ray data collection is a list of intensities, each assigned an index *hkl* corresponding to its position in the reciprocal lattice. The intensity assigned to reflection *hkl* is therefore a measure of the relative strength of the reflection from the set of lattice planes having indices *hkl*. Recall that indices are counted from the origin (indices 000), which lies in the direct path of the X-ray beam. In an undistorted image of the reciprocal lattice, such as a precession photograph (or its equivalent computed from diffractometer or oscillation data), reflections having low indices lie near the origin, and those with high indices lie farther away. Also recall that as indices increase, there is a corresponding decrease in the spacing d_{hkl} of the real-space planes represented by the indices. This means that the reflections near the origin come from sets of widely spaced planes, and thus carry information about larger features of the molecules in the unit cell. On the other hand, the reflections far from the origin come from closely spaced lattice planes in the crystal, and thus they carry information about the fine details of structure. In the next three chapters, I will examine the relationship between the intensities of the reflections and the molecular structures we seek, and thus show how the crystallographer extracts structural information from the list of intensities.

5 From Diffraction Data to Electron Density

I. Introduction

In producing an image of molecules from crystallographic data, the computer simulates the action of a lens, computing the electron density within the unit cell from the list of indexed intensities obtained by the methods described in Chapter 4. In this chapter, I will discuss the mathematical relationships between the crystallographic data and the electron density.

As I stated in Chapter 2, computation of the Fourier transform is the lens-simulating operation that a computer performs to produce an image of molecules in the crystal. The Fourier transform describes precisely the mathematical relationship between an object and its diffraction pattern. The transform allows us to convert a Fourier-series description of the reflections to a Fourier-series description of the electron density. A reflection can be described by a *structure-factor equation*, containing one term for each atom (or each volume element) in the unit cell. In turn, the electron density is described by a Fourier series in which each term is a structure factor. The crystallographer uses the Fourier transform to convert the structure factors to $\rho(x,y,z)$, the desired electron density equation.

First I will discuss Fourier series and the Fourier transform in general terms. I will emphasize the form of these equations and the information they contain, in the hope of helping you to interpret the equations—that is, to translate the equations into words and visual images. Then I will present the specific types of Fourier series that represent structure factors and electron density and show how the Fourier transform interconverts them.

II. Fourier series and the Fourier transform

A. One-dimensional waves

Recall from Chapter 2, Section VI.A, that waves are described by periodic functions, and that simple wave equations can be written in the form

$$f(x) = F\cos 2\pi(hx + \alpha) \tag{5.1}$$

or

$$f(x) = F\sin 2\pi(hx + \alpha), \tag{5.2}$$

where $f(x)$ specifies the vertical height of the wave at any horizontal position x (measured in wavelengths, where $x = 1$ implies one full wavelength or one full repeat of the periodic function). In these equations, F specifies the amplitude of the wave (its height from peak to valley), h specifies its frequency (number of wavelengths per radian), and α specifies its phase (position of the wave, in radians, with respect to the origin). These equations are *one*-dimensional in the sense that they represent a numerical value [$f(x)$, the height of the wave] at all points along *one* axis, in this case, the *x*-axis. See Fig. 2.13 for graphs of such equations.

I also stated in Chapter 2 that any wave, no matter how complicated, can be described as the sum of simple waves. This sum is called a *Fourier series* and each simple wave equation in the series is called a *Fourier term*. Either Eq. (5.1) or (5.2) could be used as a single Fourier term. For example, we can write a Fourier series of n terms using Eq. (5.1) as follows

$$f(x) = F_0\cos 2\pi(0x + \alpha_0)$$
$$+$$
$$F_1\cos 2\pi(1x + \alpha_1)$$
$$+$$
$$F_2\cos 2\pi(2x + \alpha_2) \tag{5.3}$$
$$+$$

$$\dots$$

$$+$$

$$F_n \cos 2\pi(nx + \alpha_n),$$

or equivalently,

$$f(x) = \sum_{h=0}^{n} F_h \cos 2\pi(hx + \alpha_h). \tag{5.4}$$

According to Fourier theory, any complicated periodic function can be approximated by this series, by putting the proper values of h, F_h, and α_h in each term. Think of the cosine terms as basic wave forms that can be used to build any other waveform. Also according to Fourier theory, we can use the sine function or, for that matter, *any* periodic function in the same way as the basic wave for building any other periodic function.

A very useful basic waveform is $[\cos 2\pi(hx) + i \sin 2\pi(hx)]$. Here, the waveforms of cosine and sine are combined to make a *complex number*, whose general form is $a + ib$, where i is the imaginary number $(-1)^{1/2}$. Although the phase α of this waveform is not shown, it is implicit in the combination of the cosine and sine functions, and it depends only upon the values of h and x. As I will show in Chapter 6, expressing a Fourier term in this manner gives a clear geometric means of representing the phase α and allows us to see how phases are computed. For now, just accept this convention as a convenient way to write completely general Fourier terms. In Chapter 6, I will discuss the properties of complex numbers and show how they are used to represent and compute phases.

With the terms written in this fashion, a Fourier series looks like this:

$$f(x) = \sum_{h=0}^{n} F_h [\cos 2\pi(hx) + i \sin 2\pi(hx)] \tag{5.5}$$

In words, this series is the sum of n simple Fourier terms, one for each integral value of h beginning with zero and ending with n. Each term is a simple wave with its own amplitude F_h, its own frequency h, and (implicitly) its own phase α.

Next, we can express the complex number in square brackets as an exponential, using this equality from complex number theory:

$$\cos\theta + i \sin\theta = e^{i\theta}. \tag{5.6}$$

In our case, $\theta = 2(hx)$, so the Fourier series becomes

$$f(x) = \sum_{h=0}^{n} F_h e^{2\pi i (hx)} \tag{5.7}$$

or simply

$$f(x) = \sum_h F_h e^{2\pi i(hx)}, \qquad (5.8)$$

in which the sum is taken over all values of h, and the number of terms is unspecified.

I will write Fourier series in this form throughout the remainder of the book. This kind of equation is compact and handy, but quite opaque at first encounter. Take the time now to look at this equation carefully and think about what it represents. Whenever you see an equation like this, just remember that it is a Fourier series, a sum of sine and cosine wave equations, with the full sum representing some complicated wave. The hth term in the series, $F_h e^{2\pi i(hx)}$, can be expanded to $F_h[\cos 2\pi(hx) + i \sin 2\pi(hx)]$, making plain that the hth term is a simple wave of amplitude F_h, frequency h, and implicit phase α_h.

B. Three-dimensional waves

The Fourier series that the crystallographer seeks is $\rho(x,y,z)$, the three-dimensional electron density of the molecules under study. This function is a wave equation or periodic function because it repeats itself in every unit cell. The waves described in the preceeding equations are one-dimensional: they represent a numerical value $f(x)$ that varies in one direction, along the x-axis. How do we write the equations of two-dimensional and three-dimensional waves? First, what do the graphs of such waves look like?

When you graph a function, you must use one more dimension than specified by the function. You use the additional dimension to represent the numerical value of the function. For example, in graphing $f(x)$, you use the y-axis to show the numerical value of $f(x)$. [In Fig. 2.13, the y-axes are used to represent $f(x)$, the height of each wave at point x.] Graphing a two-dimensional function $f(x,y)$ requires the third dimension to represent the numerical value of of the function.

For example, imagine a weather map with mountains whose height at location (x,y) represents the temperature at that location. Such a map graphs a two-dimensional function $t(x,y)$, which gives the temperature t at all locations (x,y) on the plane represented by the map. If we must avoid using the third dimension, for instance in order to print a flat map, the best we can do is to draw a contour map on the plane map (Fig. 5.1), with continuous lines (contours, in this case called *isotherms*) representing locations having the same temperature.

Graphing the three-dimensional function $\rho(x,y,z)$ in the same manner would require four dimensions, one for each of the spatial dimensions x, y, and z, and a fourth one for representing the value of ρ. Here a contour map is the only choice. In three dimensions, contours are continuous surfaces (rather than lines) on which the function has a constant numerical value. A contour

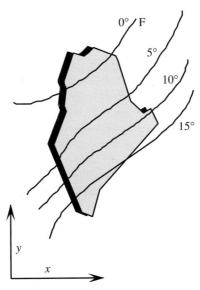

Figure 5.1   Seasonable February morning in Maine. Lines of constant temperature (isotherms) allow plotting a two-dimensional function without using the third dimension. This is a contour map of $t(x,y)$, giving the temperature t at all locations (x,y). Along each contour line lie all points having the same temperature. A planar contour map of a function of two variables takes the form of contour lines on the plane. In contrast, a contour map of a function of three variables takes the form of contour surfaces in three dimensions (see Plate 2).

map of the three-dimensional wave $\rho(x,y,z)$ exhibits surfaces of constant electron density ρ. You are already familiar with such contour maps. The common drawings of electronic orbitals (such as the 1s orbital of a hydrogen atom, often drawn as a simple sphere) is a contour map of a three-dimensional function. Everywhere on the surface of this sphere, the electron density is the same. Orbital surfaces are often drawn to enclose the region that contains 98% (or some specified value) of the total electron density.

The blue netlike surface in Plate 2 is also a contour map of a three-dimensional function. It represents a surface on which the electron density $\rho(x,y,z)$ of adipocyte lipid binding protein (ALBP) is constant. Imagine that the net encloses 98% (or some specified value) of the protein's electron density, and so the net is in essence an image of the protein's surface.

I hope the foregoing helps you to imagine three-dimensional waves. What do the equations of such waves look like? A three-dimensional wave has three frequencies, one along each of the x-, y-, and z-axes. So three variables h, k, and l are needed to specify the frequency in each of the three directions.

A general Fourier series for the wave $f(x,y,z)$, written in the compact form of Eq. (5.8) is as follows:

$$f(x, y, z) = \sum_h \sum_k \sum_l F_{hkl} e^{2\pi i(hx+k\underset{y}{+}lz)}. \tag{5.9}$$

In words, Eq. (5.9) says that the complicated three-dimensional wave $f(x,y,z)$ can be represented by a Fourier series. Each term in the series is a simple three-dimensional wave whose frequency is h in the x-direction, k in the y-direction, and l in the z-direction. For each possible set of values h, k, and l, the associated wave has amplitude F_{hkl} and, implicitly, phase α_{hkl}. The triple sum simply means to add up terms for all possible sets of integers h, k, and l. The range of values for h, k, and l depends on how many terms are required to represent the complicated wave $f(x,y,z)$ to the desired precision.

C. The Fourier transform: General features

Fourier demonstrated that for any function $f(x)$, there exists another function $F(h)$ such that

$$F(h) = \int_{-\infty}^{+\infty} f(x) e^{2\pi i(hx)} dx, \tag{5.10}$$

where $F(h)$ is called the *Fourier transform* (FT) of $f(x)$, and the units of the variable h are reciprocals of the units of x. For example, if x is time in seconds (s), then h is reciprocal time, or frequency, in reciprocal seconds (s^{-1}). So if $f(x)$ is a function of time, $F(h)$ is a function of frequency. Taking the FT of time-dependent functions is a means of decomposing these functions into their component frequencies and is sometimes referred to as *Fourier analysis*. The FT in this form is used in infrared (IR) and nuclear magnetic resonance (NMR) spectroscopy to obtain the frequencies of many spectral lines simultaneously.

On the other hand, if x is a length in Å, h is reciprocal length in $Å^{-1}$. You can thus see that this highly general mathematical form is naturally adapted for relating real and reciprocal space. In fact, as I mentioned earlier, the Fourier transform is a precise mathematical description of diffraction. The diffraction patterns in Figs. 2.07–2.10 are Fourier transforms of the corresponding simple objects and arrays. If these figures give you some intuition about how an object is related to its diffraction pattern, then they provide the same perception about the kinship between an object and its Fourier transform.

According to Eq. (5.10), to compute $F(h)$, the Fourier transform of $f(x)$, just multiply the function by $e^{2\pi i(hx)}$ and integrate (or better, let a computer integrate) the combined functions with respect to x. The result is a new function $F(h)$, which is the FT of $f(x)$. Computer programs for calculating FTs of functions are widely available.

The Fourier transform operation is reversible. That is, the same mathematical operation that gives $F(h)$ from $f(x)$ can be carried out in the opposite direction, to give $f(x)$ from $F(h)$; specifically,

$$f(x) = \int_{-\infty}^{+\infty} F(h)e^{-2\pi i(hx)}dh \qquad (5.11)$$

In other words, if $F(h)$ is the transform of $f(x)$, then $f(x)$ is in turn the transform of $F(h)$. [In this situation, $f(x)$ is sometimes called the *back-transform* of $F(h)$, but this is a loose term that simply refers to the second successive transform that recreates the original function.] Notice that the only difference between Eqs. (5.10) and (5.11) is the sign of the exponential term. You can think of this sign change as analogous (very *roughly* analogous!) to the sign change that makes subtraction the reverse of addition. Adding 3 to 5 gives 8: $5 + 3 = 8$. To reverse the operation and generate the original 5, you subtract 3 from the previous result: $8 - 3 = 5$. If you think of 8 is a simple transform of 5 made by adding 3, the back-transform of 8 is 5, produced by subtracting 3.

Returning to the visual transforms of Figs. 2.7–2.10 each object (the sphere in Fig. 2.7, for instance) is the Fourier transform (the back-transform, if you wish) of its diffraction pattern. If we build a model that looks like the diffraction pattern on the right, and then obtain its diffraction pattern, we get an image of the object on the left.

There is one added complication. The preceeding functions $f(x)$ and $F(h)$are one-dimensional. Fortunately, the Fourier transform applies to periodic functions in any number of dimensions. To restate Fourier's conclusion in three dimensions, for any function $f(x,y,z)$ there exists the function $F(h,k,l)$ such that

$$F(h,k,l) = \int_x \int_y \int_z f(x,y,z)e^{2\pi i(hx+ky+lz)} \, dx\, dy\, dz. \qquad (5.12)$$

As before, $F(h,k,l)$is called the Fourier transform of $f(x,y,z)$, and in turn, $f(x,y,z)$ is the Fourier transform of $F(h,k,l)$ as follows:

$$f(x,y,z) = \int_h \int_k \int_l F(h,k,l)e^{-2\pi i(hx+ky+lz)} \, dh\, dk\, dl. \qquad (5.13)$$

Thinking again about the potential usefulness of computing FTs in crystallography, you will see that we can use the Fourier transform to obtain information about real space, $f(x,y,z)$, from information about reciprocal space, $F(h,k,l)$. Specifically, the diffraction pattern contains information whose Fourier transform is information about the contents of the unit cell.

D. Fourier this and Fourier that: Review

I have used Fourier's name in discussing several types of equations and operations, and I want to be sure that I have not muddled them in your mind. First, a *Fourier series* is a sum of simple wave equations or periodic functions that describes or approximates a complicated periodic function. Second, constructing a Fourier series—that is, determining the proper F, h, and α values to approximate a specific function—is called *Fourier synthesis*. For example, the sum of f_0 through f_6 in Fig. 2.15 is at once a Fourier series and the product of Fourier synthesis. Third, decomposing a complicated function into its components is called *Fourier analysis*. Fourth and finally, the *Fourier transform* is an operation that transforms a function containing variables of one type (say time) into a function whose variables are reciprocals of the original type [in this case, 1/(time) or frequency]. The function $f(x)$ is related to its Fourier transform $F(h)$ by Eq. (5.10). The term *transform* is commonly used as a noun to refer to the function $F(h)$ and also loosely as a verb to denote the operation of computing a Fourier transform. (Last, there is a simple bit of grammatical awkwardness: the word *series* is both singular and plural. You must gather from context whether a writer is talking about one series or many series.)

III. Fourier mathematics and diffraction

A. Structure factor as a Fourier series

I have stated that both structure factors and electron density can be expressed as Fourier series. A structure factor describes one diffracted X-ray, which produces one reflection received at the detector. A structure factor F_{hkl} can be written as a Fourier series in which each term gives the contribution of one atom to the reflection *hkl* [see Fig. 2.15 and Eq. (2.3)]. Here is a single term, called an *atomic structure factor* f_{hkl}, in such a series, representing the contribution of the single atom j to reflection *hkl*:

$$f_{hkl} = f_j\, e^{2\pi i(hx_j + ky_j + lz_j)}. \qquad (5.14)$$

The term f_j is called the *scattering factor* of atom j, and it is a mathematical function (called a δ function) that amounts to treating the atom as a simple sphere of electron density. The function is slightly different for each element, because each element has a different number of electrons (a different value of Z) to diffract the X rays. The exponential term should be familiar to you by now. It represents a simple three-dimensional periodic function having both cosine and sine components. But the terms in parenthesis now possess added physical meaning: x_j, y_j, and z_j are the coordinates of atom j in the unit cell (real space), expressed as fractions of the unit-cell axis lengths; and h, k, and l, in addition to their role as frequencies of a wave in the three directions x, y, and z, are also the indices of a specific reflection in the reciprocal lattice.

As mentioned earlier, the phase of a wave is implicit in the exponential formulation of a structure factor and depends only upon the atomic coordinates (x_j, y_j, z_j) of the atom. In fact, the phase for diffraction by one atom is $2\pi(hx_j + ky_j + lz_j)$, the exponent of e (ignoring the imaginary i) in the structure factor. For its contribution to the 220 reflection, an atom at $(0, 1/2, 0)$ has phase $2\pi(hx_j + ky_j + lz_j)$ or $2\pi(2[0] + 2[1/2] + 0[0]) = 2\pi$, which is the same as a phase of zero. This atom lies on the (220) plane, and all atoms lying on (220) planes contribute to the 220 reflection with phase of zero. [Try the above calculation for another atom at $(1/2, 0, 0)$, which is also on a (220) plane.] This is in keeping with Bragg's law, which says that all atoms on a set of equivalent, parallel lattice planes diffract in phase with each other.

Each diffracted ray is a complicated wave, the sum of diffractive contributions from *all* atoms in the unit cell. For a unit cell containing n atoms, the structure factor F_{hkl} is the sum of all the atomic f_{hkl} values for individual atoms. Thus, in parallel with Eq. (2.3), we write the structure factor for reflection F_{hkl} as follows:

$$F_{hkl} = \sum_{j=1}^{n} f_j \, e^{2\pi i(hx_j + ky_j + lz_j)}. \tag{5.15}$$

In words, the structure factor that describes reflection hkl is a Fourier series in which each term is the contribution of one atom, treated as a simple sphere of electron density. So the contribution of each atom j to F_{hkl} depends on (1) what element it is, which determines f_j, the amplitude of the contribution, and (2) its position in the unit cell (x_j, y_j, z_j), which establishes the phase of its contribution.

Alternatively, F_{hkl} can be written as the sum of contributions from each volume element of electron density in the unit cell [see Fig. 2.16 and Eq. (2.4)]. The electron density of a volume element centered at (x,y,z) is, roughly, the average value of $\rho(x,y,z)$ in that region. The smaller we make our volume elements, the more precisely these averages approach the correct values of $\rho(x,y,z)$ at all points. We can, in effect, make our volume elements infinitesimally small, and the average values of $\rho(x,y,z)$ precisely equal to the actual values at every point, by integrating the function $\rho(x,y,z)$ rather than summing average values. Think of the resulting integral as the sum of the contributions of an infinite number of vanishingly small volume elements. Written this way,

$$F_{hkl} = \int\int\int_{h\,k\,l} \rho(x,y,z)e^{2\pi i(hx+ky+lz)}\,dx\,dy\,dz \qquad (5.16)$$

or equivalently,

$$F_{hkl} = \int_v \rho(x,y,z)e^{2\pi i(hx+ky+lz)}\,dV, \qquad (5.17)$$

where the integral over V, the unit-cell volume, is just shorthand for the integral over all values of x, y, and z in the unit cell. Each volume element contributes to F_{hkl} with a phase determined by its coordinates (x,y,z), just as the phase of atomic contributions depend on atomic coordinates.

We can see by comparing Eq. (5.17) with Eq. (5.10) [or Eq. (5.16) with Eq. (5.12)] that F_{hkl} is the Fourier transform of $\rho(x,y,z)$. More precisely, F_{hkl} is the transform of $\rho(x,y,z)$ on the set of real-lattice planes (hkl). All of the F_{hkl}s together compose the transform of $\rho(x,y,z)$ on all sets of equivalent, parallel planes throughout the unit cell.

B. Electron density as a Fourier series

Because the Fourier transform operation is reversible [Equations (5.10) and (5.11)], the electron density is in turn the transform of the structure factors, as follows:

$$\rho(x,y,z) = \frac{1}{V}\sum_h\sum_k\sum_l F_{hkl}\,e^{-2\pi i(hx+ky+lz)}, \qquad (5.18)$$

where V is the volume of the unit cell.

This transform is a triple sum rather than a triple integral because the F_{hkl}s represent a set of discrete entities: the reflections of the diffraction pattern. The transform of a discrete function, such as the reciprocal lattice of measured intensities, is a summation of discrete values of the function. The transform of a continuous function, such as $\rho(x,y,z)$, is an integral, which you can think of as a sum also, but a sum of an infinite number of infinitesimals.

Superficially, except for the sign change (in the exponential term) that accompanies the transform operation, this equation appears identical to Eq. (5.9), a general three-dimensional Fourier series. But here, each F_{hkl} is not just one of many simple numerical amplitudes for a standard set of component waves in a Fourier series. Instead, each F_{hkl} is a structure factor, itself a Fourier series, describing a specific reflection in the diffraction pattern. ("Curiouser and curiouser," said Alice.)

C. Computing electron density from data

Equation (5.18) tells us, at last, how to obtain $\rho(x,y,z)$. We need merely to construct a Fourier series from the structure factors. The structure factors describe diffracted rays that produce the measured reflections. A full description of a diffracted ray, like any description of a wave, must include three parameters: amplitude, frequency, and phase. In discussing data collection, however, I mentioned only two measurements: the indices of each reflection and its intensity. Looking again at Eq. (5.18), you see that the indices of a reflection play the role of the three frequencies in one Fourier term. The only measurable variable remaining in the equation is F_{hkl}. Does the measured intensity of a reflection, the only measurement we can make in addition to the indices, completely define F_{hkl}? Unfortunately, the answer is "no."

D. The phase problem

Because F_{hkl} is a periodic function, it possesses amplitude, frequency, and phase. It is a diffracted X ray, so its frequency is that of the X-ray source. The amplitude of F_{hkl} is proportional to the square root of the reflection intensity I_{hkl}, so structure amplitudes are directly obtainable from measured reflection intensities. But the phase of F_{hkl} is not directly obtainable from a single measurement of the reflection intensity. In order to compute $\rho(x,y,z)$ from the structure factors, we must obtain, in addition to the intensity of each reflection, the phase of each diffracted ray. In Chapter 6, I will present an expression for $\rho(x,y,z)$ as a Fourier series in which the phases are explicit, and I will discuss means of obtaining phases. This is one of the most difficult problems in crystallography. For now, on the assumption that the phases can be obtained, and thus that complete structure factors are obtainable, I will consider further the implications of Eqs. (5.15) (structure factors F expressed in terms of atoms), (5.16) [structure factors in terms of $\rho(x,y,z)$], and (5.18) [$\rho(x,y,z)$ in terms of structure factors].

IV. The meaning of the Fourier equations

A. Reflections as Fourier terms: Equation (5.18)

First consider Eq. (5.18) (ρ in terms of Fs). Each term in this Fourier-series description of $\rho(x,y,z)$ is a structure factor representing a single X-ray reflection.

The indices *hkl* of the reflection give the three frequencies necessary to describe the Fourier term as a simple wave in three dimensions. Recall from Chapter 2, Section VI.B, that any periodic function can be approximated by a Fourier series, and that the approximation improves as more terms are added to the series (see Fig. 2.14). The low-frequency terms in Eq. (5.18) determine gross features of the periodic function $\rho(x,y,z)$, whereas the high-frequency terms improve the approximation by filling in fine details. You can also see in Eq. (5.18) that the low-frequency terms in the Fourier series that describes our desired function $\rho(x,y,z)$ are given by reflections with low indices, that is, by reflections near the center of the diffraction pattern (Fig. 5.2).

The high-frequency terms are given by reflections with high indices, reflections farthest from the center of the pattern. Thus you can see the importance of how well a crystal diffracts. If a crystal does not produce diffracted rays at large angles from the direct beam (reflections with large indices), the Fourier series constructed from all the measurable reflections lacks high-frequency terms, and the resulting transform is not highly detailed—the resolution of the resulting image is poor. The Fourier series of Fig. 2.14 is truncated in just this manner and does not fit the target function in fine details like the sharp corners.

B. Computing structure factors from a model: Equations (5.15) and (5.16)

Equation (5.15) describes one structure factor in terms of diffractive contributions from all atoms in the unit cell. Equation (5.16) describes one structure factor in terms of diffractive contributions from all volume elements of electron density in the unit cell. These equations suggest that we can calculate all of the structure factors either from an atomic model of the protein or from an electron density function. In short, if we know the structure, we can calculate the diffraction pattern, including the phases of all reflections. This computation, of course, appears to go in just the opposite direction that the crystallographer desires. It turns out, however, that computing structure factors from a model of the unit cell (back-transforming the model) is an essential part of crystallography, for several reasons.

First, this computation is used in obtaining phases. As I will discuss in Chapter 6, the crystallographer obtains phases by starting from rough estimates of them and then undertaking an iterative process to improve the estimates. This iteration entails a cycle of three steps. In step 1, an estimated $\rho(x,y,z)$ (that is, a crude model of the structure) is computed using Eq. (5.18) with observed intensities (I_{obs}) and estimated phases (α_{calc}). In step 2, the crystallographer attempts to improve the model by viewing the electron-density map [a computer plot of $\rho(x,y,z)$] and identifying molecular features such as the

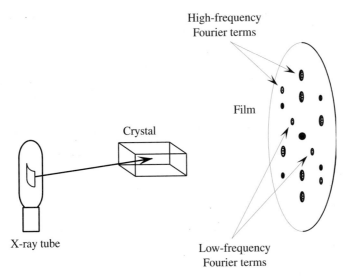

Figure 5.2 Structure factors of reflections near the center of the diffraction pattern are low-frequency terms in the Fourier series that approximates $\rho(x,y,z)$. Structure factors of reflections near the edge of the pattern are high-frequency terms.

molecule-solvent boundaries or specific groups of atoms (called interpreting the map). Step 3 entails computing new structure factors (F_{calc}), using either Eq. (5.16) with the improved $\rho(x,y,z)$ model from step 2 or Eq. (5.15) with a partial atomic model of the molecule, containing only those atoms that can be located with some confidence in the electron-density map. Calculation of new F_{calc}s in step 3 produces a new (better, we hope) set of estimated phases, and the cycle is repeated: a new $\rho(x,y,z)$ is computed from the original measured intensities and the newest phases, interpretation produces a more detailed model, and calculation of structure factors from this model produces improved phases. In each cycle, the crystallographer hopes to obtain an improved $\rho(x,y,z)$, which means a more detailed and interpretable electron-density map, and thus a more complete and accurate model of the desired structure.

I will discuss the iterative improvement of phases and electron-density maps in Chapter 7. For now just take note that obtaining the final structure entails both calculating $\rho(x,y,z)$ from structure factors and calculating structure factors from some preliminary form of $\rho(x,y,z)$. Note further that when we compute structure factors from a known or assumed model, the results include the phases. In other words, the computed results give all the information needed for a "full-color" diffraction pattern, like that shown in Plate 3d, whereas experimentally obtained diffraction patterns lack the phases and are merely black and white, like Plate 3e.

The second use of back-tranforms is to assess the progress of structure determination. Equations (5.15) and (5.16) provide means to monitor the iterative process to see whether it is converging toward improved phases and improved $\rho(x,y,z)$. The computed structure factors F_{calc} include both the desired phases α_{calc} and a new set of intensities. I will refer to these *calculated* intensities as I_{calc} to distinguish them from the *measured* reflection intensities I_{obs} taken from the diffraction pattern. As the iteration proceeds, the values of I_{calc} should approach those of I_{obs}. So the crystallographer compares the I_{calc} and I_{obs} values at each cycle in order to see whether the iteration is converging. When cycles of computation provide no further improvement in correspondence between calculated and measured intensities, then the process is complete.

C. Systematic absences in the diffraction pattern: Equation (5.15)

A third application of Eq. (5.15) allows us to understand how systematic absences in the diffraction pattern reveal symmetry elements in the unit cell, thus guiding the crystallographer in assigning the space group of the crystal. Recall from Chapter 4, Section II.H, that if the unit cell possesses symmetry elements, then certain sets of reciprocal-lattice points are equivalent, and so certain reflections in the diffraction pattern are redundant. The crystallographer must determine the unit-cell space group (i.e., determine what symmetry elements are present) in order to devise an efficient strategy for measuring as many unique reflections as possible. I stated without justification in Chapter 4 that certain symmetry elements announce themselves in the diffraction pattern as *systematic absences*: regular patterns of missing reflections. Now I will use Eq. (5.15) to show how a symmetry element in the unit cell produces systematic absences in the diffraction pattern.

For example, if the **c**-axis of the unit cell is a twofold screw axis, then reflections 001, 003, 005, along with all other 00*l* reflections in which *l* is an odd number, are missing. We can see why by using the concept of equivalent positions (Chapter 4, Section II.H). For a unit cell with a twofold screw axis along edge **c**, the equivalent positions are (x,y,z) and $(-x,-y,z+1/2)$. That is, for every atom *j* with coordinates (x,y,z) in the unit cell, there is an identical atom *j'* at $(-x,-y,z+1/2)$. Atoms *j* and *j'* are called *symmetry-related* atoms. According to Eq. (5.15), the structure factor for reflections F_{00l} is

$$F_{00l} = \sum_j f_j e^{2\pi i(lz_j)}. \tag{5.19}$$

The exponential term is greatly simplified in comparison to that in Eq. (5.15) because $h = k = 0$ for reflections on the 00l axis. Now I will separate the contributions of atoms j from that of their symmetry-related atoms j':

$$F_{00l} = \sum_j f_j e^{2\pi i (lz_j)} + \sum_{j'} f_{j'} e^{2\pi i (lz_{j'})}. \qquad (5.20)$$

Because atoms j and j' are identical, they have the same scattering factor f, and so I can substitute f_j for $f_{j'}$ and factor out the f terms:

$$F_{00l} = \sum_j f_j \left(\sum_j e^{2\pi i l z_j} + \sum_{j'} e^{2\pi i l z_{j'}} \right). \qquad (5.21)$$

If the z coordinate of atom j is z, then the z coordinate of atom j' is $z + 1/2$. Making these substitutions for z_j and $z_{j'}$,

$$F_{00l} = \sum_j f_j \left(\sum_j [e^{2\pi i l z} + e^{2\pi i l (z + 1/2)}] \right). \qquad (5.22)$$

The f_j terms are nonzero, so F_{00l} is zero, and the corresponding 00l reflection is missing only if all the summed terms in square brackets equal zero. Simplifying one of these terms,

$$e^{2\pi i l z} + e^{2\pi i l (z + 1/2)} = e^{2\pi i l z} (1 + e^{\pi i l}). \qquad (5.23)$$

This term is zero, and hence F_{00l} is zero, if $e^{\pi i l}$ equals -1. Converting this exponential to its trigonometric form (see Eq. (5.6),

$$e^{\pi i l} = \cos(\pi l) + i \sin(\pi l). \qquad (5.24)$$

The cosine of π radians (180°), or any odd multiple of π radians, is -1. The sine of π radians is 0. Thus $e^{\pi i l}$ equals -1 for all odd values of l, and F_{00l} equals zero if l is odd.

The preceding shows that F_{00l} disappears for odd values of l when the **c** edge of a unit cell is a twofold screw axis. But what is going on physically? In short, the diffracted rays from two atoms at (x, y, z) and $(-x, -y, z + 1/2)$ are identical in amplitude ($f_j = f_{j'}$) but precisely opposite in phase. Thus the pair of atoms contributes nothing to F_{00l} when l is odd. Putting it another way, if the unit cell contains a twofold screw axis along edge **c**, then every atom in the unit cell is paired with a symmetry-related atom that cancels its contributions to all odd-numbered 00l reflections.

Similar computations have been carried out for all symmetry elements and combinations of elements. Like equivalent positions, systematic absences are tabulated for all space groups in *International Tables*, so the crystallographer can use this reference as an aid to space-group determination. The *International Tables* entry for space group $P2_1$ (Fig. 4.15), which possess a 2_1 axis on edge **c**, shows that for reflections $00l$ the "Conditions limiting possible reflections" are $l = 2n$. In other words, in this space group, reflections $00l$ are present only if l is even (2 times any integer n), so they are absent if l is odd, as proved earlier.

V. Summary: From data to density

When we describe structure factors and electron density as Fourier series, we find that they are intimately related. The electron density is the Fourier transform of the structure factors, which means that we can convert the crystallographic data into an image of the unit cell and its contents. One necessary piece of information is, however, missing for each structure factor. We can measure only the intensity I_{hkl} of each reflection, not the complete structure factor F_{hkl}. What is the relationship between them? It can be shown that the amplitude of structure factor F_{hkl} is proportional to $(I_{hkl})^{1/2}$, the square root of the measured intensity. So if we know I_{hkl} from diffraction data, we know the amplitude of F_{hkl}. Unfortunately, we do not know its phase α_{hkl}. In focusing light reflected from an object, a lens maintains all phase relationships among the rays, and thus constructs an image accurately. When we record diffraction intensities, we lose the phase information that the computer needs in order to simulate an X-ray-focusing lens. In Chapter 6, I will consider how to learn the phase of each reflection, and thus to obtain the complete structure factors needed to calculate the electron density.

6

Obtaining Phases

I. Introduction

The molecular image that the crystallographer seeks is a contour map of the electron density $\rho(x,y,z)$ throughout the unit cell. The electron density, like all periodic functions, can be represented by a Fourier series. The representation that connects $\rho(x,y,z)$ to the diffraction pattern is

$$\rho(x,y,z) = \frac{1}{V}\sum_h\sum_k\sum_l F_{hkl}\, e^{-2\pi i(hx+ky+lz)}.$$

(5.18)

Equation (5.18) tells us how to calculate $\rho(x,y,z)$: simply construct a Fourier series using the structure factors F_{hkl}. For each term in the series, h, k, and l are the indices of reflection hkl, and F_{hkl} is the structure factor that describes the reflection. Each structure factor F_{hkl} is a complete description of a diffracted ray recorded as reflection hkl. Being a wave equation, F_{hkl} must specify frequency, amplitude, and phase. Its frequency is that of the X-ray source. Its amplitude is proportional to $(I_{hkl})^{1/2}$, the square root of the measured intensity I_{hkl} of reflection hkl. Its phase is unknown and is the only additional information the crystallographer needs in order to compute $\rho(x,y,z)$ and thus obtain an image of the protein. In this chapter, I will discuss some of the common methods of obtaining phases.

Let me emphasize that each reflection has a phase (see Plate 3), and so this phase problem must be solved for each one of the thousands of reflections used to construct the Fourier series that approximates $\rho(x,y,z)$. Let me also emphasize how crucial this phase information is. In his *Book of Fourier*, Kevin Cowtan illustrates the relative importance of phases and intensities in solving a structure, as shown in Plate 8. Images (*a*) and (*b*) show two simple models, a duck and a cat, along with their calculated Fourier transforms. As in Plate 3, phases are shown as colors, while the intensity of color reflects the magnitude of the structure factor at each location. (Note that these are continuous transforms, because the model is not in a lattice.) Back-transforming each Fourier transform would produce an image of the duck or cat. In (*c*) the colors from the cat transform are superimposed on the intensities from the duck transform. This gives us a transform in which the intensities come from the duck and the phases come from the cat. In (*d*) we see the back-transform of (*c*). The image of the cat is obvious, but you cannot find any sign of a duck. Ironically, the diffraction intensities, which are relatively easy to measure, contain far less information than do the phases, which are much more difficult to obtain.

In order to illuminate both the phase problem and its solution, I will represent structure factors as vectors on a two-dimensional plane of complex numbers of the form $a + ib$, where i is the imaginary number $(-1)^{1/2}$. This allows me to show geometrically how to compute phases. I will begin by introducing complex numbers and their representation as points having coordinates (a,b) on the complex plane. Then I will show how to represent structure factors as vectors on the same plane. Because we will now start thinking of the structure factor as a vector, I will hereafter write it in boldface ($\mathbf{F}_{hkl}$) instead of the italics used for simple variables and functions. Finally, I will use the vector representation of structure factors to explain a few common methods of obtaining phases.

II. Two-dimensional representation of structure factors

A. Complex numbers in two dimensions

Complex numbers of the form $N = a + ib$, where $i = (-1)^{1/2}$, can be represented as points in two dimensions (Fig. 6.1).

The horizontal axis in the figure represents the real-number line. Any real number a is a point on this line, which stretches from $a = -\infty$ to $a = +\infty$. The vertical axis is the imaginary-number line, on which lie all imaginary numbers

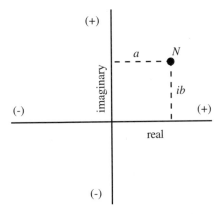

Figure 6.1 The complex number $N = a + ib$, represented as a point on the plane of complex numbers.

ib between $-i\infty$ and $+i\infty$. A complex number $a + ib$, which possesses both real (a) and imaginary (ib) parts, is thus a point at position (a,b) on this plane.

B. Structure factors as complex vectors

A representation of structure factors on this plane must include the two properties we need in order to construct $\rho(x,y,z)$: amplitude and phase. Crystallographers represent each structure factor as a *complex vector*, that is, a vector (not a point) on the plane of complex numbers. The length of this vector represents the amplitude of the structure factor. Thus the length of the vector representing structure factor $\mathbf{F}_{hkl}$ is proportional to $(I_{hkl})^{1/2}$. The second property, phase, is represented by the angle α that the vector makes with the positive real-number axis when the origin of the vector is placed at the origin of the complex plane, the point $0 + i0$ (see Fig. 6.2a).

We can represent a structure factor $\mathbf{F}$ as a vector $\mathbf{A} + i\mathbf{B}$ on this plane. The projection of $\mathbf{F}$ on the real axis is its real part $\mathbf{A}$, a vector of length $|\mathbf{A}|$ on the real-number line; and the projection of $\mathbf{F}$ on the imaginary axis is its imaginary part $i\mathbf{B}$, a vector of length $|\mathbf{B}|$ on the imaginary-number line. The length or magnitude (or in wave terminology, the amplitude) of a complex vector is analogous to the absolute value of a real number, so the length of vector $\mathbf{F}_{hkl}$ is $|\mathbf{F}_{hkl}|$; therefore, $|\mathbf{F}_{hkl}|$ is proportional to $(I_{hkl})^{1/2}$, and if the intensity is known from data collection, we can treat $|\mathbf{F}_{hkl}|$ as a known quantity. The angle that $\mathbf{F}_{hkl}$ makes with the real axis is represented in radians as α ($0 \leq \alpha \leq 2\pi$), or in cycles as α' ($0 \leq \alpha' \leq 1$), and is referred to as the *phase angle*.

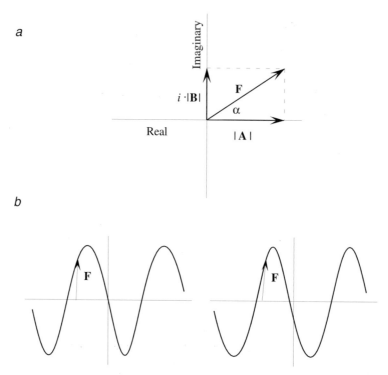

a

b

Figure 6.2   (*a*) The structure factor **F**, represented as a vector on the plane of complex numbers. The length of **F** is proportional to $I^{1/2}$, the square root of the measured intensity I, and the angle between **F** and the positive real axis is the phase α. (*b*) (Stereo pair) **F** can be pictured as a complex vector spinning around its line of travel. The projection of the path taken by the head of the vector is the familiar sine wave.

This representation of a structure factor is equivalent to thinking of a wave as a complex vector spinning around its axis as it travels thorough space (Fig. 6.2*b*). If its line of travel is perpendicular to the tail of the vector, then a projection of the head of the vector along the line of travel is the familiar sine wave. The phase of a structure factor tells us the position of the vector at some arbitrary origin, and to know the phase of all reflections means to know all their phase angles with respect to a common origin.

In Chapter 4, Section III.G, I mentioned Friedel's law, that $I_{hkl} = I_{-h-k-l}$. It will be helpful for later discussions to look at the vector representations of pairs of structure factors F_{hkl} and F_{-h-k-l}, which are called *Friedel pairs*. Even though I_{hkl} and I_{-h-k-l} are equal, $\mathbf{F}_{hkl}$ and $\mathbf{F}_{-h-k-l}$ are not. The structure factors of Friedel pairs have opposite phases, as shown in Fig. 6.3.

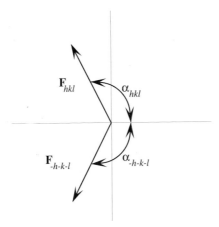

Figure 6.3   Structure factors of a Friedel pair. $\mathbf{F}_{-h-k-l}$ is the reflection of $\mathbf{F}_{hkl}$ in the real axis.

This means that $\mathbf{F}_{-h-k-l}$ is the mirror image of $\mathbf{F}_{hkl}$ with the real axis serving as the mirror. Another way to put it is that Friedel pairs are reflections of each other in the real axis.

The representation of structure factors as vectors in the complex plane (or complex vectors) is useful in several ways. Because the diffractive contributions of atoms or volume elements to a single reflection are additive, each contribution can be represented as a complex vector, and the resulting structure factor is the vector sum of all contributions. For example, in Fig. 6.4, $\mathbf{F}$ represents a structure factor of a three-atom structure, in which $\mathbf{f}_1$, $\mathbf{f}_2$, and $\mathbf{f}_3$ are the atomic structure factors.

The length of each atomic structure factor $\mathbf{f}$ represents its amplitude, and its angle α_n with the real axis represents its phase. The vector sum $\mathbf{F} = \mathbf{f}_1 + \mathbf{f}_2 + \mathbf{f}_3$ is obtained by placing the tail of $\mathbf{f}_1$ at the origin, the tail of $\mathbf{f}_2$ on the head of $\mathbf{f}_1$, and the tail of $\mathbf{f}_3$ on the head of $\mathbf{f}_2$, all the while maintaining the phase angle of each vector. The structure factor $\mathbf{F}$ is thus a vector with its tail at the origin and its head on the head of $\mathbf{f}_3$. This process sums both amplitudes and phases, so the resultant length of $\mathbf{F}$ represents its amplitude, and the resultant angle α is its phase angle. (The atomic vectors may be added in any order with the same result.)

In subsequent sections of this chapter, I will use this simple vector arithmetic to show how to compute phases from various kinds of data. In the next section, I will use complex vectors to derive an equation for electron density as a function of reflection intensities and phases.

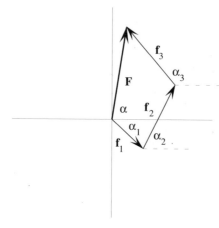

Figure 6.4 Molecular structure factor **F** is the vector sum of three atomic structure factors. Vector addition of $\mathbf{f}_1$, $\mathbf{f}_2$, and $\mathbf{f}_3$ gives the amplitude and phase of **F**.

C. Electron density as a function of intensities and phases

Figure 6.2 shows how to decompose $\mathbf{F}_{hkl}$ into its amplitude $|\mathbf{F}_{hkl}|$, which is the length of the vector, and its phase α_{hkl}, which is the angle the vector makes with the real number line. This allows us to express $\rho(x,y,z)$ as a function of the measurable amplitude of **F** (measurable because it can be computed from the reflection intensity I) and the unknown phase α. For clarity, I will at times drop the subscripts on **F**, I, and α, but remember that these relationships hold for all reflections. In Fig. 6.2,

$$\cos \alpha = \frac{|\mathbf{A}|}{|\mathbf{F}|} \quad \text{and} \quad \sin \alpha = \frac{|\mathbf{B}|}{|\mathbf{F}|} \tag{6.1}$$

and therefore

$$|\mathbf{A}| = |\mathbf{F}| \cdot \cos \alpha \quad \text{and} \quad |\mathbf{B}| = |\mathbf{F}| \cdot \sin \alpha. \tag{6.2}$$

Expressing **F** as a complex vector $\mathbf{A} + i\mathbf{B}$,

$$\mathbf{F} = |\mathbf{A}| + i|\mathbf{B}| = |\mathbf{F}| \cdot (\cos\alpha + i\sin\alpha). \tag{6.3}$$

Expressing the complex term in the parentheses as an exponential (Eq. (5.6)),

$$\mathbf{F} = |\mathbf{F}| \cdot e^{i\alpha}. \tag{6.4}$$

Substituting this expression for F_{hkl} in Eq. 5.18, the electron-density equation (remembering that α is the phase α_{hkl} of a specific reflection), gives

$$\rho(x,y,z) = \frac{1}{V} \sum_h \sum_k \sum_l |F_{hkl}|\, e^{i\alpha_{hkl}} e^{-2\pi i(hx+ky+lz)}.$$

$$(6.5)$$

We can combine the exponential terms more simply by expressing the phase angle as α', using $\alpha = 2\pi\alpha'$:

$$\rho(x,y,z) = \frac{1}{V} \sum_h \sum_k \sum_l |F_{hkl}|\, e^{2\pi i\alpha'_{hkl}} e^{-2\pi i(hx+ky+lz)}.$$

$$(6.6)$$

Now we can combine the exponentials by adding their exponents

$$\rho(x,y,z) = \frac{1}{V} \sum_h \sum_k \sum_l |F_{hkl}|\, e^{-2\pi i(hx+ky+lz-\alpha'_{hkl})}.$$

$$(6.7)$$

This equation gives the desired electron density as a function of the known amplitudes $|F|$ and the unknown phases α'_{hkl} of each reflection. Recall that this equation represents $\rho(x,y,z)$ in a now-familiar form, as a Fourier series, but this time with the phase of each structure factor expressed explicitly. Each term in the series is a three-dimensional wave of amplitude $|F_{hkl}|$, phase α'_{hkl}, and frequencies h along the x-axis, k along the y-axis, and l along the z-axis.

The most demanding element of macromolecular crystallography (except, perhaps, for dealing with macromolecules that resist crystallization) is the so-called phase problem, that of determining the phase angle α_{hkl} for each reflection. In the remainder of this chapter, I will discuss some of the common methods for overcoming this obstacle. These include the *heavy-atom method* (also called *isomorphous replacement*), *anomalous scattering* (also called *anomalous dispersion*), and *molecular replacement*. Each of these techniques yield only estimates of phases, which must be improved before an interpretable electron-density map can be obtained. In addition, these techniques usually yield estimates for a limited number of the phases, so phase determination must be extended to include as many reflections as possible. In Chapter 7, I will discuss methods of phase improvement and phase extension, which ultimately result in accurate phases and an interpretable electron-density map.

III. The heavy-atom method (isomorphous replacement)

Each atom in the unit cell contributes to every reflection in the diffraction pattern [Eq. (5.15)]. The contribution of an atom is greatest to the reflections whose indices correspond to lattice planes that intersect that atom, so a specific atom contributes to some reflections strongly, and to some weakly or not at all. If we could add one or a very small number of atoms to identical sites in all unit cells of a crystal, we would expect to see changes in the diffraction pattern, as the result of the additional contributions of the added atom. As I will show later, the slight perturbation in the diffraction pattern caused by an added atom can be used to obtain initial estimates of phases. In order for these perturbations to be large enough to measure, the added atom must be a strong diffractor, which means it must be an element of high atomic number, a so-called heavy atom.

A. Preparing heavy-atom derivatives

After obtaining a complete set of X-ray data and determining that these data are adequate to produce a high-resolution structure, the crystallographer undertakes to prepare one or more *heavy-atom derivatives*. In the most common technique, crystals of the protein are soaked in solutions of heavy ions, for instance ions or ionic complexes of Hg, Pt, or Au. In many cases, such ions bind to one or a few specific sites on the protein without perturbing its conformation or crystal packing. For instance, surface cysteine residues react readily with Hg^{2+} ions, and cysteine, histidine, and methionine displace chloride from Pt complexes like $PtCl_4^{2-}$ to form stable Pt adducts. The conditions that give such specific binding must be found by simply trying different ionic compounds at various pH values and concentrations.

Several diffraction criteria define a promising heavy-atom derivative. First, the derivative crystals must be *isomorphic* with native crystals. At the molecular level, this means that the heavy atom must not disturb crystal packing or the conformation of the protein. Unit-cell dimensions are quite sensitive to such disturbances, so heavy-atom derivatives whose unit-cell dimensions are the same as native crystals are probably isomorphous. The term *isomorphous replacement* comes from this criterion.

The second criterion for useful heavy-atom derivatives is that there must be measurable changes in at least a modest number of reflection intensities. These changes are the handle by which phase estimates are pulled from the data, so they must be clearly detectable, and large enough to measure accurately.

Fig. 6.5 shows precession photographs for native and derivative crystals of the MoFe protein of nitrogenase. Underlined in the figure are pairs of reflections whose relative intensities are altered by the heavy atom. In examining heavy-atom photos by eye, the crystallographer looks for pairs of reflections whose relative intensities are reversed. This distinguishes real heavy-atom perturbations from simple differences in overall intensity of two photos. For example, consider the leftmost underlined pairs in each photograph. In the native photo (*a*), the reflection on the right is the darker of the pair, whereas in

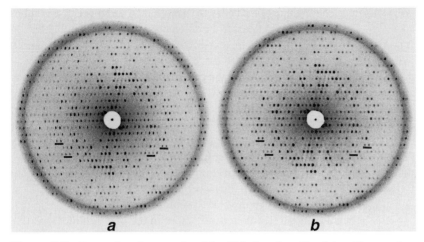

Figure 6.5   Precession photographs of the $hk0$ plane in native (a) and heavy-atom (b) crystals of the MoFe protein from nitrogenase. Corresponding underlined pairs in the native and heavy-atom patterns show reversed relative intensities. Photos courtesy of Professor Jeffrey Bolin.

the derivative photo (b), the reflection on the left is darker. Several additional differences suggest that this derivative might produce good phases.

Finally, the derivative crystal must diffract to reasonably high resolution, although the resolution of derivative data need not be as high as that of native data. Methods of phase extension (Chapter 7) can produce phases for higher-angle reflections from good phases of reflections at lower angles.

Having obtained a suitable derivative, the crystallographer faces data collection again. Because derivatives must be isomorphous with native crystals, the strategy is the same as that for collecting native data. You can see that the phase problem effectively multiplies the magnitude of the crystallographic project by the number of derivative data sets needed. As I will show, at least two, and often more, derivatives are required.

B. Obtaining phases from heavy-atom data

Consider a single reflection of amplitude $|\mathbf{F_P}|$ (P for protein) in the native data, and the corresponding reflection of amplitude $|\mathbf{F_{HP}}|$ (HP for heavy atom plus protein) in data from a heavy-atom derivative. Because the diffractive contributions of all atoms to a reflection are additive, the difference in amplitudes ($|\mathbf{F_{HP}}| - |\mathbf{F_P}|$)

is the amplitude contribution of the heavy atom alone, and the square of this difference, $(|\mathbf{F}_{HP}| - |\mathbf{F}_{P}|)^2$, is proportional to the difference $I_{HP}-I_{P}$. (Remember that $|\mathbf{F}|$ is proportional to $I^{1/2}$.) If we compute a diffraction pattern in which the amplitude of each reflection is $(|\mathbf{F}_{HP}| - |\mathbf{F}_{P}|)^2$, the result is the diffraction pattern of the heavy atom alone in the protein's unit cell. In effect, we have subtracted away all contributions from the protein atoms, leaving only the heavy-atom contributions. Now we see the diffraction pattern of one (or only a small number) of atoms, rather than the far more complex pattern of the protein.

In comparison to the protein structure, this "structure"—a sphere (or very few spheres) in a lattice—is very simple. It us usually easy to "determine" this structure, that is, to find the location of the heavy atom in the unit cell. Before considering how to locate the heavy atom (Section III.C.), I will show how finding it helps us to solve the phase problem.

Suppose we are able to locate a heavy atom in the unit cell of derivative crystals. Recall that Eq. (5.15) gives us the means to calculate the structure factors $\mathbf{F}_{hkl}$ for a known structure. This calculation gives us not just the amplitudes but the complete structure factors, including each of their phases. So we can compute the amplitudes and phases of our simple structure, the heavy atom in the protein unit cell. Now consider a single reflection hkl as it appears in the native and derivative data. Let the structure factor of the native reflection be $\mathbf{F}_{P}$. Let the structure factor of the corresponding derivative reflection be $\mathbf{F}_{HP}$. Finally, let $\mathbf{F}_{H}$ be the structure factor for the heavy atom itself, which we can compute if we can locate the heavy atom.

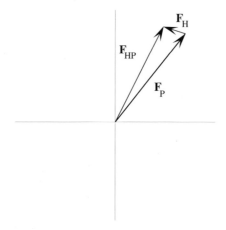

Figure 6.6   A structure factor $\mathbf{F}_{HP}$ for the heavy-atom derivative is the sum of contributions from the native structure ($\mathbf{F}_{P}$) and the heavy atom ($\mathbf{F}_{H}$).

Figure 6.6 shows the relationship among the vectors $\mathbf{F}_H$, $\mathbf{F}_{HP}$, and $\mathbf{F}_H$ on the complex plane. (Remember that we are considering this relationship for a specific reflection, but the same relationship holds for all reflections.) Because the diffractive contributions of atoms are additive vectors,

$$\mathbf{F}_{HP} = \mathbf{F}_H + \mathbf{F}_P. \qquad (6.8)$$

That is, the structure factor for the heavy-atom derivative is the vector sum of the structure factors for the protein alone and the heavy atom alone.

For each reflection, we wish to know $\mathbf{F}_P$. (We already know that its length is obtainable from the measured reflection intensity I_P, but we want to learn its phase angle.) According to the previous equation,

$$\mathbf{F}_P = \mathbf{F}_{HP} - \mathbf{F}_H. \qquad (6.9)$$

We can solve this vector equation for $\mathbf{F}_P$, and thus obtain the phase angle of the structure factor, by representing the equation in the complex plane (Fig. 6.7).

We know $|\mathbf{F}_{HP}|$ and $|\mathbf{F}_P|$ from measuring reflection intensities I_{HP} and I_P. So we know the length of the vectors $\mathbf{F}_{HP}$ and $\mathbf{F}_P$, but not their directions or phase angles. We know $\mathbf{F}_H$, including its phase angle, from locating the heavy atom and calculating all its structure factors. To solve Eq. (6.9) for $\mathbf{F}_H$ and

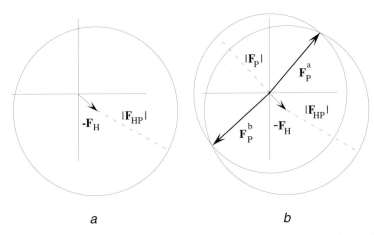

a   b

Figure 6.7 Vector solution of Equation (6.9). (*a*) All points on the circle equal the vector sum $|\mathbf{F}_{HP}| - \mathbf{F}_H$. (*b*) Vectors from the origin to intersections of the two circles are solutions to Eq. (6.9).

thus obtain its phase angle, we place the vector $-\mathbf{F}_H$ at the origin and draw a circle of radius $|\mathbf{F}_{HP}|$ centered on the head of vector $-\mathbf{F}_H$ (Fig. 6.7a). All points on this circle equal the vector sum $|\mathbf{F}_{HP}|$ $-\mathbf{F}_H$. In other words, we know that the head of $\mathbf{F}_{HP}$ lies somewhere on this circle of radius $|\mathbf{F}_{HP}|$. Next we add a circle of radius $|\mathbf{F}_P|$ centered at the origin (Fig. 6.7b). We know that the head of the vector $\mathbf{F}_P$ lies somewhere on this circle, but we do not know where because we do not know its phase angle. Equation (6.9) holds only at points where the two circles intersect. Thus the phase angles of the two vectors $\mathbf{F}_P^a$ and $\mathbf{F}_P^b$ that terminate at the points of intersection of the circles are the only possible phases for this reflection.

Our heavy-atom derivative allows us to determine, for each reflection hkl, that α_{hkl} has one of two values. How do we decide which of the two phases is correct? In some cases, if the two intersections lie near each other, the average of the two phase angles will serve as a reasonable estimate. I will show in Chapter 7 that certain phase improvement methods can sometimes succeed with such phases from only one derivative, in which case the structure is said to be solved by the method of *single isomorphous replacement* (SIR). More commonly, however, a second heavy-atom derivative must be found and the vector problem outlined previously must be solved again. Of the two possible phase angles found by using the second derivative, one should agree better with one of the two solutions from the first derivative, as shown in Fig. 6.8.

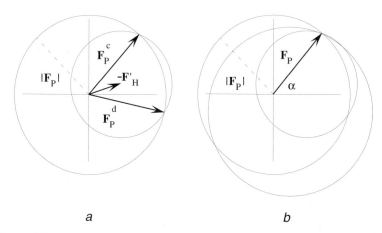

a b

Figure 6.8   (a) A second heavy-atom derivative indicates two possible phases, one of which corresponds to $\mathbf{F}_a$ in Fig. 6.7b. (b) $\mathbf{F}_P$, which points from the origin to the common intersection of the three circles, is the solution to Equation (6.9) for both heavy-atom derivatives. Thus α is the correct phase for this reflection.

Figure 6.8a shows the phase determination using a second heavy-atom derivative; $\mathbf{F}'_H$ is the structure factor for the second heavy atom. The radius of the smaller circle is $|\mathbf{F}'_{HP}|$, the amplitude of $\mathbf{F}'_{HP}$ for the second heavy-atom derivative. For this derivative, $\mathbf{F}_P = \mathbf{F}'_{HP} - \mathbf{F}'_H$. Construction as before shows that the phase angles of $\mathbf{F}^c_P$ and $\mathbf{F}^d_P$ are possible phases for this reflection. In Fig. 6.8b, the circles from Figs. 6.7b and 6.8a are superimposed, showing that $\mathbf{F}^c_P$ is identical to $\mathbf{F}^a_P$. This common solution to the two vector equations is $\mathbf{F}_P$, the desired structure factor. The phase of this reflection is therefore the angle labeled α in the figure, the only phase compatible with data from both derivatives.

In order to resolve the phase ambiguity from the first heavy-atom derivative, the second heavy atom must bind at a different site from the first. If two heavy atoms bind at the same site, the phases of $\mathbf{F}_H$ will be the same in both cases, and both phase determinations will provide the same information. This is true because the phase of an atomic structure factor depends only on the location of the atom in the unit cell, and not on its identity (Chapter 5, Section III.A). In practice, it sometimes takes three or more heavy-atom derivatives to produce enough phase estimates to make the needed initial dent in the phase problem. Obtaining phases with two or more derivatives is called the method of multiple isomorphous replacement (MIR). This is the method by which most protein structures have been determined.

To compute a high-resolution structure, we must ultimately know the phases α_{hkl} for all reflections. High-speed computers can solve large numbers of these vector problems rapidly, yielding an estimate of each phase along with a measure of its precision.[1] For many phases, the precision of the first phase estimate is so low that the phase is unusable. For instance, in Fig. 6.9, the circles graze each other rather than intersecting sharply, so there is a large uncertainty in α. In some cases, because of inevitable experimental errors in measuring intensities, the circles do not intersect at all.

Computer programs for calculating phases also compute statistical parameters representing attempts to judge the quality of phases. Some parameters, usually called *phase probabilities*, are measures of the uncertainty of individual phases. Others parameters, including figure of merit, closure errors, phase differences, and various R-factors are attempts to assess the quality of groups of phases obtained by averaging results from several heavy-atom derivatives (or results from other phasing methods). In most cases, these parameters are numbers between 0 (poor phases) and 1 (perfect phases). No single one of

---

[1]Computer programs calculate phases for each derivative numerically (rather than geometrically) by obtaining two solutions to the equation

$$\mathbf{F}^2_{HP} = \mathbf{F}^2_P + \mathbf{F}^2_H + 2\mathbf{F}_P\mathbf{F}_H \cos(\alpha_P + \alpha_H).$$

The pairs of solutions for two heavy-atom derivatives should have one solution in common.

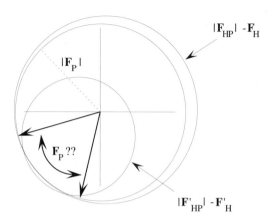

Figure 6.9   The MIR solution for this structure factor gives phase of high uncertainty.

these statistics is an accurate measure of the goodness of phases. Crystal-lographers often use two or more of these criteria simultaneously in order to cull out questionable phases. In short, until correct phases are obtained (see Chapter 7), there is no sure way to measure the quality of estimates. The acid test of phases is whether they give an interpretable electron-density map.

When promising phases are available, the crystallographer carries out Fourier synthesis (Eq. (6.7)) to calculate $\rho(x,y,z)$. Each Fourier term is multi-plied by the probability of correctness of the associated phase. This procedure gives greater weight to terms with more reliable phases. Every phase that de-fies solution or is too uncertain (and for that matter every intensity that is too weak to measure accurately) forces the crystallographer to omit one term from the Fourier series when calculating $\rho(x,y,z)$. Each omitted term lowers the ac-curacy of the approximation to $\rho(x,y,z)$, degrading the quality and resolution of the resulting map. In practice, a good pair of heavy-atom derivatives may allow us to estimate only a small percentage of the phases. We can enlarge our list of precise phases by iterative processes mentioned briefly in Chapter 5, Section IV.B, which I will describe more fully in Chapter 7. For now, I will complete this discussion of the heavy-atom method by considering how to find heavy atoms, which is necessary for calculation of $\mathbf{F}_H$.

C. Locating heavy atoms in the unit cell

Before we can obtain phase estimates by the method described in the previous section, we must locate the heavy atoms in the unit cell of derivative crystals.

As I described earlier, this entails extracting the relatively simple diffraction signature of the heavy atom from the far more complicated diffraction pattern of the heavy-atom derivative, and then solving a simpler "structure," that of one heavy atom (or a few) in the unit cell of the protein. The most powerful tool in determining the heavy-atom coordinates is a Fourier series called the *Patterson function* $P(u,v,w)$, a variation on the Fourier series used to compute $\rho(x,y,z)$ from structure factors. The coordinates (u,v,w) locate a point in a *Patterson map*, in the same way that coordinates (x,y,z) locate a point in an electron-density map. The Patterson function or Patterson synthesis is a Fourier series without phases. The amplitude of each term is the square of one structure factor, which is proportional to the measured reflection intensity. Thus we can construct this series from intensity measurements, even though we have no phase information. Here is the Patterson function in general form

$$P(u,v,w) = \frac{1}{V} \sum_h \sum_k \sum_l |\mathbf{F}^2_{hkl}| \, e^{-2\pi i(hu+kv+lw)}.$$

$$(6.10)$$

To obtain the Patterson function solely for the heavy atoms in derivative crystals, we construct a *difference Patterson function*, in which the amplitudes are $(\Delta\mathbf{F})^2 = (|\mathbf{F}_{HP}| - |\mathbf{F}_P|)^2$. The difference between the structure-factor amplitudes with and without the heavy atom reflects the contribution of the heavy atom alone. The difference Patterson function is

$$\Delta P(u,v,w) = \frac{1}{V} \sum_h \sum_k \sum_l \Delta\mathbf{F}^2_{hkl} \, e^{-2\pi i(hu+kv+lw)}.$$

$$(6.11)$$

In words, the difference Patterson function is a Fourier series of simple sine and cosine terms. (Remember that the exponential term is shorthand for these trigonometric functions.) Each term in the series is derived from one reflection hkl in both the native and derivative data sets, and the amplitude of each term is $(|\mathbf{F}_{HP}| - |\mathbf{F}_P|)^2$, which is the amplitude contribution of the heavy atom to structure factor $\mathbf{F}_{HP}$. Each term has three frequencies h in the u-direction, k in the v-direction, and l in the w-direction. Phases of the structure factors are not included; at this point, they are unknown.

Because the Patterson function contains no phases, it can be computed from any raw set of crystallographic data, but what does it tell us? A contour map of $\rho(x,y,z)$ displays areas of high density (peaks) at the locations of atoms. In contrast, a Patterson map, which is a contour map of $P(u,v,w)$, displays peaks at locations corresponding to *vectors between atoms*. (This is a strange idea at first, but the following example will make it clearer.) Of course, there are more vectors between atoms than there are atoms, so a Patterson map is more complicated than an electron-density map. But if the structure is simple, like that of one or a few heavy atoms in the unit cell, the Patterson map may be simple enough to allow us to locate the atom(s). You can see now that the main reason for using the difference Patterson function instead of a simple Patterson using F_{HP}s is to eliminate the enormous number of peaks representing vectors between light atoms in the protein.

I will show, in a two-dimensional example, how to construct the Patterson map from a simple crystal structure and then how to use a calculated Patterson map to deduce a structure (Fig. 6.10). The simple molecular structure in Fig. 6.10a contains three atoms (dark circles) in each unit cell. To construct the Patterson map, first draw all possible vectors between atoms in one unit cell, including vectors between the same pair of atoms but in opposite directions. (For example, treat $1 \rightarrow 2$ and $2 \rightarrow 1$ as distinct vectors.) Two of the six vectors ($1 \rightarrow 3$ and $3 \rightarrow 2$) are shown in the figure. Then draw empty unit cells around an origin (Fig. 6.10b), and redraw all vectors with their tails at the origin. The head of each vector is the location of a peak in the Patterson map, sometimes called a Patterson "atom" (light circles). The coordinates (u,v,w)

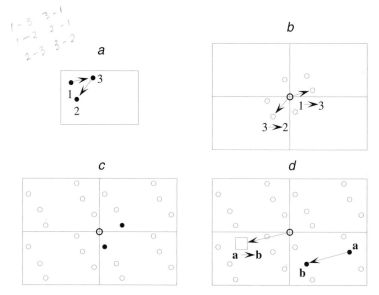

Figure 6.10 Construction and interpretation of a Patterson map. (a) Structure of unit cell containing three atoms. Two of the six interatomic vectors are shown. (b) Patterson map is constructed by moving all interatomic vectors to the origin. Patterson "atoms" (peaks in the contour map) occur at the head of each vector. (c) Complete Patterson map, containing all peaks from (b) in all unit cells. Peak at origin results from self-vectors. Image of original structure is present (origin and two darkened peaks) amid other peaks. (d) Trial solution of map (c). If origin and Patterson atoms **a** and **b** were the image of the real unit cell, the interatomic vector $\mathbf{a} \rightarrow \mathbf{b}$ would produce a peak in the small box. Absence of the peak disproves this trial solution.

of a Patterson atom representing a vector between atom 1 at (x_1,y_1,z_1) and atom 2 at (x_2,y_2,z_2) are $(u,v,w) = (x_1 - x_2, y_1 - y_2, z_1 - z_2)$. The vectors from Fig. 6.10a are redrawn in Fig. 6.10b, along with all additional Patterson atoms produced by this procedure. Finally, in each of the unit cells, duplicate the Patterson atoms from all four unit cells. The result (Fig. 6.10c) is a complete Patterson map of the structure in Fig. 6.10a. In this case, there are six Patterson atoms in each unit cell. You can easily prove to yourself that a real unit cell containing n atoms will give a Patterson unit cell containing $n(n-1)$ Patterson atoms.

Now let's think about how to go from a computed Patterson map to a structure—that is, how to locate real atoms from Patterson atoms. A computed Patterson map exhibits a strong peak at the origin because this is the location of all vectors between an atom and itself. Notice in Fig. 6.10c that the origin and two of the Patterson atoms (dark circles) reconstruct the original arrangement of atoms in Fig. 6.10a. Finding six peaks (ignoring the peak at the origin) in each unit cell of the calculated Patterson map, we infer that there are three real atoms per unit cell. [Solve the equation $n(n-1) = 6$.] We therefore know that the origin and two peaks reconstruct the relationship among the three real atoms, but we do not know which two peaks to choose. To solve the problem, we pick a set of peaks—the origin and two others—as a trial solution, and follow the rules described earlier to generate the expected Patterson map for this arrangement of atoms. If the trial map has the same peaks as the calculated map, then the trial arrangement of atoms is correct. By trial and error, we can determine which pair of Patterson atoms, along with an atom at the origin, would produce the remaining Patterson atoms. Figure 6.10d shows an incorrect solution (the origin plus peaks **a** and **b**). The vector **a** → **b** is redrawn at the origin to show that the map does not contain the Patterson atom **a** → **b**, and hence that this solution is incorrect.

You can see that as the number of real atoms increases, the number of Patterson atoms, and with it the difficulty of this problem, increases rapidly. Computer programs can search for solutions to such problems and, upon finding a solution, can refine the atom positions to give the most likely arrangement of heavy atoms.

Unit-cell symmetry can also simplify the search for peaks in a three-dimensional Patterson map. For instance, in a unit cell with a 2_1 axis (twofold screw) on edge **c**, recall (equivalent positions, Chapter 4, Section II.H) that each atom at (x,y,z) has an identical counterpart atom at $(-x,-y,1/2 + z)$. The vectors connecting such symmetry-related atoms will all lie at $(u,v,w) = (2x,2y,1/2)$ in the Patterson map (just subtract one set of coordinates from the other), which means they all lie in the plane that cuts the Patterson unit cell at $w = 1/2$. Such planes, which contain the Patterson vectors for symmetry-related atoms, are called *Harker sections* or *Harker planes*. If heavy atoms bind to the protein at equivalent positions, heavy-atom peaks in the Patterson map can be found on the Harker sections. (Certain symmetry elements give Patterson vectors that all lie upon a line, called *a Harker line*, rather than on a plane.)

There is an added complication: the arrangement of heavy atoms in a protein unit cell is often enantiomeric. For example, if heavy atoms are found along a threefold screw axis, the screw may be left- or right-handed. The Patterson map does not distinguish between mirror image, or more accurately, inverted-image, arrangements of heavy atoms because you cannot tell whether a Patterson vector **ab** is **a** → **b** or **b** → **a**. But the phases obtained by calculating structure factors from the inverted model are incorrect and will not lead to an interpretable map. Crystallographers refer to this difficulty as the *hand problem* (although hands are mirror images and the two solutions described here are inversions). If derivative data are available to high resolution, the crystallographer simply calculates two electron-density maps, one with phases from each enantiomer of the heavy-atom structure. With luck, one of these maps will be distinctly clearer than the other. If derivative data are only available at low resolution, this method may not determine the hand with certainty. The problem may require the use of anomalous scattering methods, discussed in Section IV.F.

Having located the heavy atom(s) in the unit cell, the crystallographer can compute the structure factors F_H for the heavy atoms alone, using Eq. (5.15). This calculation yields both the amplitudes and the phases of structure factors F_H, giving the vector quantities needed to solve Eq. (6.9) for the phases α_{hkl} of protein structure factors F_P. This completes the information needed to compute a first electron-density map, using Eq. (6.7). This map requires improvement because these first phase estimates contain substantial errors. I will discuss improvement of phases and maps in Chapter 7.

IV. Anomalous scattering

A. Introduction

A second means of obtaining phases from heavy-atom derivatives takes advantage of the heavy atom's capacity to absorb X rays of specified wavelength. As a result of this absorption, Friedel's law (Chapter 4, Section III.G) does not hold, and the reflections hkl and $-h-k-l$ are not equal in intensity. This inequality of symmetry-related reflections is called *anomalous scattering* or *anomalous dispersion*.

Recall from Chapter 4, Section III.B that elements *absorb* X rays as well as emit them, and that this absorption drops sharply at wavelengths just below their characteristic emission wavelength K_β (Fig. 4.16). This sudden change in absorption as a function of λ is called an *absorption edge*. An element exhibits anomalous scattering when the X-ray wavelength is near the element's absorption edge. Absorption edges for the light atoms in the unit cell are not near the wavelength of X rays used in crystallography, so carbon, nitrogen, and oxygen do not contribute to anomalous scattering. However, absorption edges of heavy atoms are in this range, and if X rays of varying wavelength are available, as is often the case at synchrotron sources, X-ray data can be collected under conditions that maximize anomalous scattering by the heavy atom.

B. The measurable effects of anomalous scattering

When the X-ray wavelength is near the heavy-atom absorption edge, a fraction of the radiation is absorbed by the heavy atom and re-emitted with altered phase. The effect of this anomalous scattering on a given structure factor $\mathbf{F}_{HP}$ in the heavy-atom data is depicted in vector diagrams as consisting of two perpendicular contributions, one real ($\mathbf{\Delta F}_r$) and the other imaginary ($\mathbf{\Delta F}_i$).

In Fig. 6.11, $\mathbf{F}_{HP}^{\lambda 1}$ represents a structure factor for the heavy atom derivative measured at wavelength λ_1, where anomalous scattering does not occur. $\mathbf{F}_{HP}^{\lambda 2}$ is the same structure factor measured at a second X-ray wavelength λ_2 near the absorption edge of the heavy atom, so anomalous scattering alters the heavy-atom contribution to this structure factor. The vectors representing

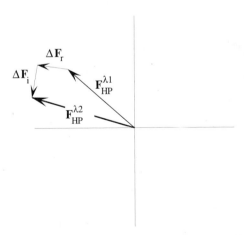

Figure 6.11   Real and imaginary anomalous-scattering contributions alter the magnitude and phase of the structure factor.

anomalous scattering contributions are $\Delta\mathbf{F}_r$ (real) and $\Delta\mathbf{F}_i$ (imaginary). From the diagram, you can see that

$$\mathbf{F}_{HP}^{\lambda 2} = \mathbf{F}_{HP}^{\lambda 1} + \Delta\mathbf{F}_r + \Delta\mathbf{F}_i. \tag{6.12}$$

Fig. 6.12 shows the result of anomalous scattering for a Friedel pair of structure factors, distinguished from each other in the figure by superscripts $+$ and $-$. Recall that for Friedel pairs in the absence of anomalous scattering, $|\mathbf{F}_{hkl}| = |\mathbf{F}_{-h-k-l}|$ and $\alpha_{hkl} = -\alpha_{-h-k-l}$, so $\mathbf{F}_{HP}^{\lambda 1-}$ is the reflection of $\mathbf{F}_{HP}^{\lambda 1+}$ in the real axis. The real contributions $\Delta\mathbf{F}_r^+$ and $\Delta\mathbf{F}_r^-$ to the reflections of a Friedel pair are, like the structure factors themselves, reflections of each other in the real axis. On the other hand, it can be shown (but I will not prove it here) that the imaginary contribution to $\mathbf{F}_{HP}^{\lambda 1-}$ is the inverted reflection of that for $\mathbf{F}_{HP}^{\lambda 1+}$. That is, $\Delta\mathbf{F}_i^-$ is obtained by reflecting $\Delta\mathbf{F}_i^+$ in the real axis and then reversing its sign or pointing it in the opposite direction. Because of this difference between the imaginary contributions to these reflections, under anomalous scattering the two structure factors are no longer precisely equal in intensity, nor are they precisely opposite in phase. It is clear from Fig. 6.12 that $\mathbf{F}_{HP}^{\lambda 2-}$ is not the mirror image of $\mathbf{F}_{HP}^{\lambda 2+}$. From this disparity between Friedel pairs, the crystallographer can extract phase information.

C. Extracting phases from anomalous scattering data

The magnitude of anomalous scattering contributions $\Delta\mathbf{F}_r$ and $\Delta\mathbf{F}_i$ for a given element are constant and roughly independent of reflection angle θ, so these quantities can be looked up in tables of crystallographic information. The phases of $\Delta\mathbf{F}_r$ and $\Delta\mathbf{F}_i$ depend only upon the position of the heavy atom in the unit cell, so once the heavy atom is located by Patterson methods, the phases can be computed. The resulting full knowledge of $\Delta\mathbf{F}_r$ and $\Delta\mathbf{F}_i$ allows Eq. (6.12) to be solved for the vector $\mathbf{F}_{HP}^{\lambda 1}$, thus establishing its phase. Crystallographers obtain solutions by computer, but I will solve the general equation using complex vector diagrams (Fig. 6.13), and thus show that the amount of information is adequate to solve the problem.

First consider the structure factor $\mathbf{F}_{HP}^{\lambda 1+}$ in Fig. 6.12. Applying Eq. (6.12) and solving for $\mathbf{F}_{HP}^{\lambda 1+}$ gives

$$\mathbf{F}_{HP}^{\lambda 1+} = \mathbf{F}_{HP}^{\lambda 2+} - \Delta\mathbf{F}_r^+ - \Delta\mathbf{F}_i^+. \tag{6.13}$$

To solve this equation (see Fig. 6.13), draw the vector $-\Delta\mathbf{F}_r^+$ with its tail at the origin, and draw $-\Delta\mathbf{F}_i^+$ with its tail on the head of $-\Delta\mathbf{F}_r^+$. With the

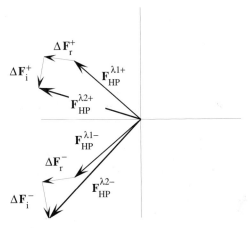

Figure 6.12   Under anomalous scattering, $\mathbf{F}_{-h-k-l}$ is no longer the mirror image of $\mathbf{F}_{hkl}$.

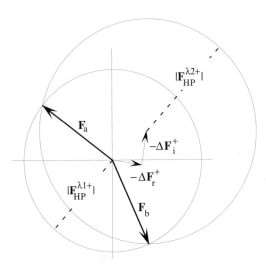

Figure 6.13   Vector solution of Eq. (6.13). $\Delta\mathbf{F}_r$ and $\Delta\mathbf{F}_i$ play the same role as $\mathbf{F}_H$ in Figs. 6.7 and 6.8.

head of $-\Delta\mathbf{F}_i^+$ as center, draw a circle of radius $|\mathbf{F}_{HP}^{\lambda2}{}^+|$, representing the amplitude of this reflection in the anomalous scattering data set. The head of the vector $\mathbf{F}_{HP}^{\lambda2}{}^+$ lies somewhere on this circle. We do not know where because we do no know the phase of the reflection. Now draw a circle of radius

$|F_{HP}^{\lambda 1 +}|$ with its center at the origin, representing the structure-factor amplitude of this same reflection in the nonanomalous scattering data set. The two
points of intersection of these circles satisfy Eq. (6.13), establishing the phase
of this reflection as either that of $\mathbf{F_a}$ or $\mathbf{F_b}$. As with the SIR method, we cannot
tell which of the two phases is correct.

The Friedel partner of this reflection comes to the rescue. We can obtain
a second vector equation involving $F_{HP}^{\lambda 1 +}$ by reflecting $F_{HP}^{\lambda 2 -}$ and all its
vector components across the real axis (Fig. 6.14a).

After reflection, $F_{HP}^{\lambda 1 -}$ equals $F_{HP}^{\lambda 1 +}$, ΔF_r^- equals ΔF_r^+, and ΔF_i^- equals
ΔF_i^+. The magnitude of $F_{HP}^{\lambda 2 -}$ is unaltered by reflection across the real axis.
If we make these substitutions in Eq. (6.13), we obtain

$$F_{HP}^{\lambda 1+} = |F_{HP}^{\lambda 2-}| - \Delta F_r^+ - \Delta F_i^+. \tag{6.14}$$

We can solve this equation in the same manner that we solved Eq. (6.13),
by placing the vectors $-\Delta F_r^+$ and ΔF_i^+ head-to-tail at the origin, and draw-
ing a circle of radius $|F_{HP}^{\lambda 2 -}|$ centered on the head of $+\Delta F_i^+$ (Fig. 6.14b).
Finally, we draw a circle of radius $|F_{HP}^{\lambda 1 +}|$ centered at the origin. The circles

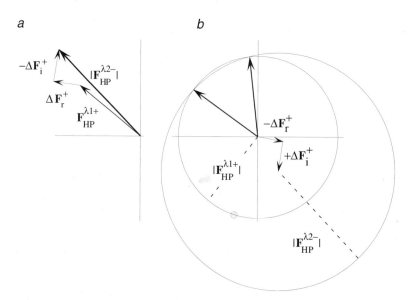

Figure 6.14   Reflection of $\mathbf{F}^-$ components across the real axis gives a second vector
equation involving the desired structure factor. (a) All reflected components are
labeled with their equivalent contributions from $\mathbf{F}^+$. (b) Vector solution of Eq. (6.14).
These solutions are compatible only with $\mathbf{F_a}$ in Fig. 6.13.

intersect at the two solutions to Eq. (6.14). Although the circles graze each other and give two phases with considerable uncertainty, one of the possible solutions corresponds to $\mathbf{F}_a$ in Fig. 6.13, and neither of them is close to the phase of $\mathbf{F}_b$.

So the disparity between intensities of Friedel pairs in the anomalous scattering data set establishes their phases in the *non*anomalous scattering data set. The reflection whose phase has been established here corresponds to the vector $\mathbf{F}_{HP}$ in Eq. (6.9). Thus the amplitudes and phases of two of the three vectors in the Eq. (6.9) are known: (1) $\mathbf{F}_{HP}$ is known from the anomalous scattering computation just shown, and (2) $\mathbf{F}_H$ is known from calculating the heavy-atom structure factors after locating the heavy atom by Patterson methods. The vector $\mathbf{F}_P$, then, is simply the vector difference $\mathbf{F}_{HP} - \mathbf{F}_H$, establishing the phase of this reflection in the native data.

D. Summary

Under anomalous scattering, the members of a Friedel pair can be used to establish the phase of a reflection in the heavy atom derivative data, thus establishing the phase of the corresponding reflection in the native data. Let me briefly review the entire project of obtaining the initial structure factors by SIR with anomalous scattering. First, we collect a complete data set with native crystals, giving us the amplitudes $|\mathbf{F}_{Phkl}|$ for each native reflection. Then we find a heavy-atom derivative and collect a second data set at the same wavelength, giving amplitudes $|\mathbf{F}_{HPhkl}|$ for each reflection in the heavy-atom data. Next we collect a third data set at a different X-ray wavelength, chosen to maximize anomalous scattering by the heavy atom. We use the nonequivalence of Friedel pairs in the anomalous scattering data to establish phases of reflections in the heavy-atom data, and we use the phased heavy-atom derivative structure factors to establish the native phases. (Puff puff!)

In practice, several of the most commonly used heavy atoms (including uranium, mercury, and platinum) give strong anomalous scattering with Cu–K_α radiation. In such cases, crystallographers can measure intensities of Friedel pairs in the heavy-atom data set. In phase determination (refer to Figs. 6.12–14), the average of $|\mathbf{F}_{hkl}|$ and $|\mathbf{F}_{-h-k-l}|$ serves as both $|\mathbf{F}_{HP}^{\lambda 1 \; +}|$ and $|\mathbf{F}_{HP}^{\lambda 1 \; -}|$, while $|\mathbf{F}_{hkl}|$ and $|\mathbf{F}_{-h-k-l}|$ separately serve as $|\mathbf{F}_{HP}^{\lambda 2 \; +}|$ and $|\mathbf{F}_{HP}^{\lambda 2 \; -}|$, so only one heavy atom data set is required.

Like phases from the MIR method, each anomalous scattering phase can only serve as an initial estimate and must be weighted with some measure of phase probability. The intensity differences between Friedel pairs are very small, so measured intensities must be very accurate if any usable phase information is to be derived. To improve accuracy, the crystallographer collects intensities of

Friedel partners under very similar conditions, and always from the same crystal. Diffractometry is ideal for anomalous scattering because of its inherently great accuracy in measuring intensities, and because the diffractometer can be programmed to collect Friedel pairs in succession, thus assuring that the crystal is in the same condition during collection of the two reflections. But diffractometry is slow compared to collecting data with area detectors.

E. Multiwavelength anomalous diffraction phasing

Three developments—variable-wavelength synchrotron X rays, cryocrystallography, and the production of proteins containing selenomethionine instead of the normal sulfur-containing methionine—have recently allowed rapid progress in maximizing the information obtainable from anomalous dispersion. For proteins that naturally contain a heavy atom, such as the iron in a globin or cytochrome, the native heavy atom provides the source of anomalous dispersion. Proteins lacking functional heavy atoms can be expressed in *Escherichia coli* containing exclusively selenomethionine. The selenium atoms then serve as heavy atoms in a protein that is essentially identical to the "native" form. Isomorphism is, of course, not a problem with these proteins because the same protein serves as both the native and derivative forms.

The power of multiwavelength radiation is that data sets from a heavy-atom derivative at different wavelengths are in many respects like those from distinct heavy-atom derivatives. Especially in the neighborhood of the absorption maximum of the heavy atom [see, for example, the absorption spectra of copper and nickel (dotted lines) in Fig. 4.16], the real and imaginary anomalous scattering factors ΔF_r and ΔF_i vary greatly with X-ray wavelength. At the absorption maximum, ΔF_i reaches its maximum value, whereas at the ascending inflection point or edge, ΔF_r reaches a minimum and then increases farther from the absorption peak. So data sets taken at the heavy-atom absorption maximum, the edge, and at wavelengths distance from the maximum all have distinct values for the real and imaginary contributions of anomalous dispersion. Thus each measurement of a Freidel pair at a specific wavelength provides the components of distinct sets of phasing equations like those solved in Figs. 6.13 and 6.14. In addition to wavelength-dependent differences between Friedel pairs, individual reflection intensities vary slightly with wavelength (called dispersive differences), and these differences also contain phase information, which can be extracted by solving equations much like those for isomorphous replacement. All told, data sets at different wavelengths from a single crystal can contain sufficient phasing information to solve a structure if the molecule under study contains one or more atoms that give anomalous dispersion. This method is called multiwavelength anomalous dispersion, or MAD, phasing.

The principles upon which MAD phasing are based have been known for years. But the method had to await the availability of variable-wavelength synchroton X-ray sources. In addition, the intensity differences the crystallographer must measure are small, and until recently, these small signals were difficult to measure with sufficient precision. To maximize accuracy in the measured differences between Friedel pairs, corresponding pairs should be measured at nearly the same time so that crystal condition and instrument parameters do not change between measurements. And even more demanding, crystal condition and instrument parameters should also be as constant as possible for complete data sets taken at different wavelengths. These technical demands are met by cryocrystallography and synchrotron sources. Flash-freezing preserves the crystal in essentially unchanged condition through extensive data collection. Synchrotron sources provide X rays of precisely controllable wavelength and also of high intensity, which shortens collection time.

The first successes of MAD phasing were small proteins that contain functional heavy atoms. Production of selenomethionine proteins opened the door to MAD phasing for nonmetalloproteins. But larger proteins may contain too many methionines, and thus too many heavy atoms. Because more heavy atoms should mean stronger anomalous dispersion, how can there be too many? Recall that to solve SIR and anomalous-dispersion phase equations, we must know the position(s) of the heavy atom(s) in the unit cell. For MAD phasing, heavy atoms can be located by Patterson methods (Section III.C), which entails trial-and-error comparisons of the Patterson map with calculated Pattersons for various proposed models of heavy-atom locations. But Patterson maps of proteins containing many heavy atoms may require so many trial solutions that they are impossible to solve. More sophisticated methods of locating the heavy atom(s), entailing least-squares fitting of the data from all wavelengths to find the heavy atom structure factors, can sometimes work for larger numbers of heavy atoms. As of this writing, the largest proteins solved by "conventional" MAD phasing (but see Section IV.G) was sulfite reductase, with a molecular mass of 64,000, 497 residues, and five Fe atoms as anomalous scatterers. The largest number of Se atoms in a structure solved by MAD was six per protein subunit, in the enzyme DsbA, a dimeric disulfide oxidoreductase with subunit mass of 21,000 and 189 residues per subunit.

F. Anomalous scattering and the hand problem

As I discussed in Section III.C, Patterson methods do not allow us to distinguish between enantiomeric arrangements of heavy atoms, and phases derived from heavy-atom positions of the wrong hand are incorrect. When high-resolution data are available for the heavy-atom derivative, phases and electron-density maps can be calculated for both enantiomeric possibilities. The map calculated with phases from the correct enantiomer will sometimes be demonstrably sharper and more interpretable. If not, and if anomalous scattering data are available, SIR *and* anomalous scattering phases can be computed for both hands, and maps can be prepared from the two sets of phases. The added phase information from anomalous scattering sometimes makes hand selection possible when SIR phases alone do not.

The availability of two heavy-atom derivatives, one with anomalous scattering, allows a powerful technique for establishing the hand, even at quite low resolution. Heavy atoms in the first derivative are located by Patterson methods, one of the possible hands is chosen, and SIR phases are computed. Then, using the same hand assumption, we compute anomalous-scattering phases. For the second heavy-atom derivative, instead of using Patterson methods, we compute a difference Fourier between the native data and the second derivative data, using the SIR phases from the first derivative. Then we compute a second difference Fourier, adding the phases from anomalous scattering. Finally, we compute a third difference Fourier, just like the second except that the signs of all anomalous-scattering contributions are reversed, which is like assuming the opposite hand. The first Fourier should exhibit electron-density peaks at the positions of the second heavy atom. If the initial hand assumption was correct, heavy-atom peaks should be stronger in the second Fourier. If it was incorrect, heavy-atom peaks should be stronger in the third Fourier.

G. Direct phasing: Application of methods from small-molecule crystallography

Methods involving heavy atoms apply almost exclusively to large molecules (500 or more atoms, not counting hydrogens). For small molecules (up to 200 atoms), phases can be determined by what are commonly called *direct methods*. One form of direct methods relies on the existence of mathematical relationships among certain combinations of phases. From these relationships, a sufficient number of initial phase estimates can be obtained to begin converging toward a complete set of phases. Another form, executed by a program called Shake-and-Bake, in essence tries out random arrangements of atoms, simulates the diffraction patterns they would produce, and compares the simulated patterns with those obtained from the crystals. Even though the trial arrangements are limited to those that are physically possible (for example, having no two atoms closer than bonding or van der Waals forces allow), the number of trial arrangements can be too large for computation if the number of atoms is large.

Direct methods work if the molecules, and thus the unit cells and numbers of reflections, are relatively small. Isomorphous replacement works if the molecules are large enough that a heavy atom does not disturb their structures significantly. The most difficult structures for crystallographers are those that are too large for direct methods and too small to remain isomorphous despite the intrusion of a heavy atom. If a medium-size protein naturally contains a heavy atom, like iron or zinc, or if a selenomethionine derivative can be produced, the structure can often be solved by MAD phasing (Section IV.E). [NMR methods (see Chapter 10) are also of great power for small and medium-size molecules.]

The Shake-and-Bake style of direct phasing apparently has the potential to solve the structures of proteins of over 100 residues if they diffract exceptionally well (to around 1.0 Å). Fewer than 10% of large molecules diffract well enough to qualify. But in a combined process that shows great promise, direct phasing has been combined with MAD phasing to solve a few large structures. Recall that larger proteins may contain too many methionines to allow Patterson or least-squares location of all seleniums in the selenomethionine derivative. Solving the anomalous-diffraction phase equations requires knowing the locations of all the heavy atoms. Shake-and-Bake can solve this problem, even if there are 50 or more seleniums in the protein. One of the first successes of this combined method was protein of molecular mass over 250,000 containing 65 selenomethionines.

Our last phasing method applies to all molecules, regardless of size, but it requires knowledge that the desired structure is similar to a known structure.

V. Molecular replacement: Related proteins as phasing models

A. Introduction

The crystallographer can sometimes use the phases from structure factors of a known protein as initial estimates of phases for a new protein. If this method is feasible, then the crystallographer may be able to determine the structure of the new protein from a single native data set. The known protein in this case is referred to as a *phasing model*, and the method, which entails calculating initial phases by placing a model of the known protein in the unit cell of the new protein, is called *molecular replacement*.

For instance, the mammalian serine proteases—trypsin, chymotrypsin, and elastase—are very similar in structure and conformation. If a new mammalian serine protease is discovered, and sequence homology with known proteases suggests that this new protease is similar in structure to known ones, then one of the known proteases might be used as a phasing model for determining the structure of the new protein.

Similarly, having learned the crystallographic structure of a protein, we may want to study the conformational changes that occur when the protein binds to a small ligand and to learn the molecular details of protein-ligand binding. We might be able to crystallize the protein and ligand together or introduce the ligand into protein crystals by soaking. We expect that the protein/ligand complex is similar in structure to the free protein. If this expectation is realized, we do not have to work completely from scratch to determine the structure of the complex. We can use the ligand-free protein as a phasing model for the protein–ligand complex.

In Plate 8, I showed that phases contain more information than intensities. How, then, can the phases from a different protein help us find an unknown structure? In his *Book of Fourier*, Kevin Cowtan uses computed transforms to illustrate this concept, as shown in Plate 9. First we see, posing this time as an unknown structure, the cat (*a*), with its Fourier transform shown in black and white. The colorless transform is analogous to an experimental diffraction pattern because we do not observe phases in experimental data. Next we see a Manx (tailless) cat (*b*) (along with its transform) posing as a solved structure that we also know (because of, say, sequence homology) to be similar in structure to the cat (*a*). If we know that the unknown structure, the cat, is similar to a known structure, the Manx cat, are the intensities of the cat powerful enough to reveal the differences between the unknown structure and the phasing model—in this case, the tail? In (*c*) the phases (colors) of the Manx cat transform are superimposed on the intensities of the unknown cat transform. In (*d*) we see the back-transform of (*c*), and although the image is weak, the cat's tail is apparent. The intensities of cat diffraction do indeed provide enough information to show how the cat differs from the Manx cat. In like manner, measured intensities from a protein of unknown structure do indeed have the power to show us how it differs from a similar, known structure used as a phasing model.

B. Isomorphous phasing models

If the phasing model and the new protein are isomorphous, as may be the case when a small ligand is soaked into protein crystals, then the phases from the free protein can be used directly to compute $\rho(x,y,z)$ from native intensities of the new protein [Eq. (6.15)].

$$\rho(x,y,z) = \frac{1}{V} \sum_h \sum_k \sum_l |\mathbf{F}_{hkl}^{new}| \, e^{-2\pi i(hx+ky+lz-\alpha'^{model}_{hkl})}.$$

(6.15)

In this Fourier synthesis, the amplitudes $|\mathbf{F}_{hkl}^{new}|$ are obtained from the native intensities of the new protein, and the phases α'^{model} are those of the phasing model. During the iterative process of phase improvement (Chapter 7), the phases should change from those of the model to those of the new protein or complex, revealing the desired structure. In Plate 9, we not only knew that our phasing model was similar to the unknown, but we had the added advantage of knowing that its orientation was the same. Otherwise, its phases would not have revealed the unknown structure.

C. Nonisomorphous phasing models

If the phasing model is not isomorphous with the desired structure, the problem is more difficult. The phases of atomic structure factors, and hence of molecular structure factors, depend upon the location of atoms in the unit cell. In order to use a known protein as a phasing model, we must superimpose the structure of the model on the structure of the new protein in its unit cell and then calculate phases for the properly oriented model. In other words, we must find the position and orientation of the phasing model in the new unit cell that would give phases most like those of the new protein. Then we can calculate the structure factors of the properly positioned model and use the phases of these computed structure factors as initial estimates of the desired phases.

Without knowing the structure of the new protein, how can we copy the model into the unit cell with the proper orientation and position? From native data on the new protein, we can determine its unit-cell dimensions and symmetry. Clearly the phasing model must be placed in the unit cell with the same symmetry as the desired protein. This places some constraints upon where to place the model, but not enough to give useful estimates of phases. In theory, it should be possible to conduct a computer search of all orientations and positions of the model in the new unit cell. For each trial position and orientation, we would calculate the structure factors (called $\mathbf{F}_{calc}$) of the model [Eq. (5.15)], and compare their amplitudes $|\mathbf{F}_{calc}|$ with the measured amplitudes $|\mathbf{F}_{obs}|$ obtained from diffraction intensities of the new protein. Finding the position and orientation that gives the best match, we would take the computed phases (α_{calc}) as the starting phases for structure determination of the new protein.

D. Separate searches for orientation and location

In practice, the number of trial orientations and positions for the phasing model is enormous, so a brute-force search is impractical, even on the fastest computers. The procedure is greatly simplified by separating the search for the best orientation from the search for the best position. Further, it is possible to search for the best orientation independently of location by using the Patterson function.

If you consider the procedure for drawing a Patterson map from a known structure (Section III.C), you will see that the final map is independent of the position of the structure in the unit cell. No matter where you draw the "molecule," as long as you do not change its orientation (that is, as long as you do not rotate it within the unit cell), the Patterson map looks the same. On the other hand, if you rotate the structure in the unit cell, the Patterson map rotates around the origin, altering the arrangement of Patterson atoms in a single Patterson unit cell. This suggests that the Patterson map might provide a means of determining the best orientation of the model in the unit cell of the new protein.

If the model and the new protein are indeed similar, and if they are oriented in the same way in unit cells of the same dimensions and symmetry, they should give very similar Patterson maps. We might imagine a trial-and-error method in which we compute Patterson maps for various model orientations and compare them with the Patterson map of the desired protein. In this manner, we could find the best orientation of the model, and then use that single orientation in our search for the best position of the model, using the structure-factor approach outlined earlier.

How much computing do we actually save by searching for orientation and location separately? The orientation of the model can be specified by three angles of rotation about orthogonal axes x, y, and z with their origins at the center of the model. Specifying location also requires three numbers, the x, y, and z coordinates of the molecular center with respect to the origin of the unit cell. For the sake of argument, let us say that we must try 100 different values for each of the six parameters. (In real situations, the number of trial values is much larger.) The number of combinations of six parameters, each with 100 possible values is 100^6, or 10^{12}. Finding the orientation as a separate search requires first trying 100 different values for each of three angles, which is 100^3 or 10^6 combinations. After finding the orientation, finding the location requires trying 100 different values of each of three coordinates, again 100^3 or 10^6 combinations. The total number of trials for separate orientation and location searches is $10^6 + 10^6$ or 2×10^6. The magnitude of the saving is $10^{12}/2\times10^6$ or 500,000. In this case, the problem of finding the orientation and location separately is smaller by half a million times than the problem of searching for orientation and location simultaneously.

E. Monitoring the search

Finally, what mathematical criteria are used in these searches? In other words, as the computer goes through sets of trial values (angles or coordinates) for the model, how does it compare results and determine optimum values of the parameters?

For the orientation search (often called a *rotation search*), the computer is looking for large values of the model Patterson function $P^{\text{model}}(u,v,w)$ at locations corresponding to peaks in the Patterson map of the desired protein. A powerful and sensitive way to evaluate the model Patterson is to compute the minimum value of $P^{\text{model}}(u,v,w)$ at all locations of peaks in the Patterson map of the desired protein. A value of zero for this minimum means that the trial orientation has no peak in at least one location where the desired protein exhibits a peak. A high value for this minimum means that the trial orientation has peaks at all locations of peaks in the Patterson map of the desired protein.

For the location search, the criterion is the correspondence between the expected structure-factor amplitudes from the model in a given trial location and the actual amplitudes derived from the native data on the desired protein. This criterion can be expressed as the R-factor, a parameter we will encounter later as a criterion of improvement of phases in final structure determination. The R-factor compares overall agreement between the amplitudes of two sets of structure factors, as follows

$$R = \frac{\sum \left| |\mathbf{F}_{\text{obs}}| - |\mathbf{F}_{\text{calc}}| \right|}{\sum |\mathbf{F}_{\text{obs}}|}. \tag{6.16}$$

In words, for each reflection, we compute the difference between the observed structure-factor amplitude from the native data set $|\mathbf{F}_{\text{obs}}|$ and the calculated amplitude from the model in its current trial location $|\mathbf{F}_{\text{calc}}|$ and take the absolute value, giving the magnitude of the difference. We add these magnitudes for all reflections. Then we divide by the sum of the observed structure-factor amplitudes (the reflection intensities).

If, on the whole, the observed and calculated intensities agree with each other, the differences in the numerator are small, and the sum of the differences is small compared to the sum of the intensities themselves, so R is small. For perfect agreement, all the differences equal zero, and R equals zero. No single difference is likely to be larger than the corresponding $|\mathbf{F}_{\text{obs}}|$, so the maximum value of R is one. For proteins, R-values of 0.3 to 0.4 for the best placement of a phasing model have often provided adequate initial estimates of phases.

F. Summary

If we know that the structure of a new protein is similar to that of a known protein, we can use the known protein as a phasing model, and thus solve the phase problem without heavy atom derivatives. If the new crystals and those of the model are isomorphous, the model phases can be used directly as estimates of the desired phases. If not, we must somehow superimpose the known protein upon the new protein to create the best phasing model. We can do this without knowledge of the structure of the new protein by using Patterson-map comparisons to find the best orientation of the model protein and then using structure-factor comparisons to find the best location of the model protein.

VI. Iterative improvement of phases (preview of Chapter 7)

The phase problem greatly increases the effort required to obtain an interpretable electron-density map. In this chapter, I have discussed several methods of obtaining phases. In all cases, the phases obtained are estimates, and often the set of estimates is incomplete. Electron-density maps calculated from Eq. (6.7), using measured amplitudes and first phase estimates, are often difficult or impossible to interpret. In Chapter 7, I will discuss improvement of phase estimates and extension of phase assignments to as many reflections as possible. As phase improvement and extension proceed, electron-density maps become clearer and easier to interpret as an image of a molecular model. The iterative process of *structure refinement* eventually leads to a structure that is in good agreement with the original data.

7

Obtaining and Judging
the Molecular Model

I. Introduction

In this chapter, I will discuss the final stages of structure determination: obtaining and improving the electron-density map, interpreting the map to produce an atomic model of the unit-cell contents, and refining the model to optimize its agreement with the original native reflection intensities. The criteria by which the crystallographer judges the progress of the work overlap with criteria for assessing the quality of the final model. These subjects form the bridge from Chapter 7 to Chapter 8, where I will review many of the concepts of this book by guiding you through the experimental descriptions from the description of a structure determination as published in a scientific journal.

II. Iterative improvement of maps and models: Overview

In brief, obtaining a detailed molecular model of the unit-cell contents entails calculating $\rho(x,y,z)$ from Eq. (6.7) using measured intensities from the native data set and phases computed from heavy-atom data, anomalous scattering, or molecular replacement. Because the phases are rough estimates, the first map may be uninformative and disappointing. Crystallographers improve the map by an iterative process sometimes called bootstrapping. The basic principle of this iteration is easy to state but demands care, judgment, and much labor to execute: any features that can be reliably discerned in the map become part of a phasing model for subsequent maps.

Whatever crude model of unit cell contents that can be discerned in the map is cast in the form of a simple electron-density function and used to calculate new structure factors by Eq. (5.16). The phases of these structure factors are used, along with the original native intensities, to add more terms to Eq. (6.7), the Fourier-series description of $\rho(x,y,z)$, in hopes of producing a clearer map. When the map becomes clear enough to allow location of atoms, these are added to the model, and structure factors are computed from this model using Eq. (5.15), which contains atomic structure factors rather than electron density. As the model becomes more detailed, the phases computed from it improve, and the model, computed from the original native structure-factor amplitudes and the latest phases, becomes even more detailed. The crystallographer thus tries to bootstrap from the initial rough phase estimates to phases of high accuracy, and from them, a clear, interpretable map and a model that fits the map well.

I should emphasize that the crystallographer cannot get any new phase information without modifying the model in some way. The possible modifications include solvent flattening, noncrystallographic symmetry averaging, or introducing a partial atomic model, all of which are discussed further in this chapter. It is considered best practice to make sure that intial phases are good enough to make the map interpretable. If it is not, then the crystallographer needs to find additional derivatives and collect better data.

The model can be improved in another way: by least-squares refinement of the atomic coordinates. This method entails adjusting the atomic coordinates to improve the agreement between amplitudes calculated from the current model and the original measured amplitudes in the native data set. In the latter stages of structure determination, the crystallographer alternates between map interpretation and least-squares refinement.

The block diagram in Fig. 7.1 shows how these various methods ultimately produce a molecular model that agrees with the native data. The vertical dotted line in Fig. 7.1 divides the operations into two categories. To the right of the the line are real-space methods, which entail attempts to improve the electron-density map, by adding information to the map or removing noise from it, or to improve the model, using the map as a guide. To the left of the line are reciprocal-space methods, which entail attempts to improve phases or to improve the agreement between reflection intensities computed from the

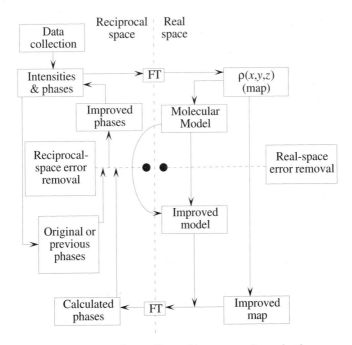

Figure 7.1 Block diagram of crystallographic structure determination.

model and the original measured reflection intensities. In real-space methods, the criteria for improvement or removal of errors are found in electron-density maps, in the fit of model to map, or in the adherence of the model to expected bond lengths and angles (all real-space criteria); in reciprocal-space methods, the criteria for improvement or removal of errors involve reliability of phases and agreement of calculated structure factors with measured intensities (all reciprocal-space criteria). The link between real and reciprocal space is, of course, the Fourier transform (FT).

I will return to this diagram near the end of the chapter, particularly to amplify the meaning of *error removal,* which is indicated by dashed horizontal lines in Fig. 7.1. For now, I will illustrate the bootstrapping technique for improving phases, map, and model with an analogy: the method of successive approximations for solving a complicated algebraic equation. Most mathematics education emphasizes equations that can be solved analytically for specific variables. Many realistic problems defy such analytic solutions but are amenable to numerical methods. The method of successive approximations has much in common with the iterative process that extracts a protein model from diffraction data.

Consider the problem of solving the following equation for the variable y.

$$\left(1 + \frac{1}{y^2}\right)(y - 1) = 0. \tag{7.1}$$

Attempts to simplify the equation produce a cubic equation in y, giving no straightforward means to a numerical solution. You can, however, easily obtain a numerical solution for y with a hand calculator. Start by solving for y in terms of y^2 as follows

$$y = \frac{1}{\left(1 + \frac{1}{y^2}\right)} + 1. \tag{7.2}$$

Then make an arbitrary initial estimate of y, say $y = 1$. (This is analogous to starting with the MIR phases as initial estimates of the correct phases.) Plug this estimate into the right-hand y^2 term, and calculate y [analogous to computing a crude structure from measured structure-factor amplitudes ($|\mathbf{F}_{obs}|$) and phase estimates]. The result is 1.5. Now take this computed result as the next estimate (analogous to computing new structure factors from the crude structure), plug it into the y^2 term, and compute y again (analogous to computing a new structure from better phase estimates). The result is 1.6923. Repeating this process produces these answers in succession 1.7412, 1.752, 1.7543, 1.7547, 1.7549, 1.7549, and so on. After a few iterations, the process converges to a solution; that is, the output value of y is the same as the input. This value is a solution to the original equation.

With Eq. (7.2), any first estimate above 1.0 (even one million) produces the result shown. In contrast, for many other equations, the method of successive approximations works only if the initial estimate is close to a correct solution. Otherwise, the successive answers do not converge; instead, they may oscillate among several values (the iteration "hangs up" instead of converging), or they may continually become larger in magnitude (the iteration "blows up"). In order for the far more complex crystallographic iteration to converge to a protein model that is consistent with the diffraction data, initial estimates of many phases must be close to the correct values. Attempts to start from random phases in hopes of convergence to correct ones appear doomed to failure because of the large number of incorrect solutions to which the process can converge.

The following sections describe the crystallographic bootstrapping process in more detail.

III. First maps

A. Resources for the first map

Entering the final stages of structure determination, the crystallographer is armed with several sets of data with which to calculate $\rho(x,y,z)$ as a Fourier series of structure factors using Eq. (5.18). First is the original native data set, which usually contains the most accurate and complete (highest-resolution) set of measured intensities. These data will support the most critical tests of the final molecular model. Next are data sets from heavy-atom derivatives, which are often limited to lower resolution. Several sets of phases may be available, calculated from heavy-atom derivatives and perhaps anomalous dispersion. Because each phase must be calculated from a heavy-atom reflection, phase estimates are not available for native reflections at higher resolution than that of the best heavy-atom derivative. Finally, for each set of phases, there is usually some criterion of precision. These criteria will be used as weighting factors, numbers between 0 and 1, for Fourier terms containing the phases. A Fourier term containing a phase estimate of low reliability will be multiplied by a low weighting factor in the Fourier-series computation of $\rho(x,y,z)$. In other words, such a term will be multiplied by a number less than 1.0 to reduce its contribution to the Fourier series, and thus reduce bias from a reflection whose phase is questionable. Conversely, a term containing a phase of high reliability will be given full weight (weighting factor of 1.0) in the series.

Here is the Fourier series that gives the first electron-density map

$$\rho(x,y,z) = \frac{1}{V}\sum_{h}\sum_{k}\sum_{l} w_{hkl}|\mathbf{F}_{obs}|\,e^{-2\pi i(hx+ky+lz-\alpha'_{calc})}. \tag{7.3}$$

In words, the desired electron-density function is a Fourier series in which term hkl has amplitude $|\mathbf{F}_{obs}|$, which equals $(I_{hkl})^{1/2}$, the square root of the measured intensity I_{hkl} from the native data set. The phase α'_{hkl} of the same term is calculated from heavy-atom, anomalous dispersion, or molecular replacement data, as described in Chapter 6. The term is weighted by the factor w_{hkl}, which will be near 1.0 if α'_{hkl} is among the most highly reliable phases, or smaller if the phase is questionable. This Fourier series is called an $\mathbf{F}_{obs}$ or $\mathbf{F}_{o}$ synthesis (and the map an $\mathbf{F}_{o}$ map) because the amplitude of each term hkl is $|\mathbf{F}_{obs}|$ for reflection hkl.

The first term in this Fourier series, the $\mathbf{F}_{000}$ term, should contain $(I_{000})^{1/2}$, where I_{000} is the intensity of reflection 000, which lies at the origin of the reciprocal lattice. Recall that this reflection is never measured because it is obscured by the direct beam. Examination of Eq. (7.3) reveals that $\mathbf{F}_{000}$ is a real constant (as opposed to a complex or imaginary number). The phase α'_{000} of this term is assigned a value of zero, with the result that all other phases will be computed relative to this assignment. Then because $h = k = l = 0$ for reflection 000, the exponent of e is zero and the entire exponential term is 1.0. Thus F_{000} is a constant, just like f_0 in Fig. 2.14.

All other terms in the series are simple trigonometric functions with average values of zero, so it is clear that the value assigned to F_{000} will determine the overall amplitude of the electron-density map. (In the same manner, the f_0 term in Fig. 2.14 displaces all the Fourier sums upward, making the sums positive for all values of x, like the target function.) The sensible assignment for F_{000} is therefore the total number of electrons in the unit cell, making the sum of $\rho(x,y,z)$ over the whole unit cell equal to the total electron density. In practice, this term can be omitted from the calculation, and the overall map amplitude can be set by means described in Section III.C.

B. Displaying and examining the map

Until the mid to late 1980s, the contour map of the first calculated electron density was displayed by printing sections of the unit cell onto Plexiglas or clear plastic sheets and stacking them to produce a three-dimensional model, called a *minimap*. Today's computers can display the equivalent of a minimap and allow much more informative first glimpses of the electron density. These glimpses of the molecular image are often attended with great excitement and expectation. If the phase estimates are sufficiently good, the map will show some of the gross features of unit cell contents. In the rare best cases, with good phases from molecular replacement, and perhaps with enhancement from noncrystallographic averaging (explained further in Section III.C.) first maps are easily interpretable, clearly showing continuous chains of electron density and features like alpha helices—perhaps even allowing some amino-acid side chains to be identified. At the worst, the first map is singularly uninformative, signaling the need for additional phasing information, perhaps from another heavy-atom derivative. Usually the minimum result that promises a structure from the existing data is that protein be distinguishable from bulk water, and that dense features like alpha helices can be recognized. If the boundary of each molecule, the *molecular envelope*, shows some evidence of recognizable protein structure, then a full structure is likely to come forth.

I will consider the latter case, in which the first map defines a molecular envelope, with perhaps a little additional detail. If more detail can be discerned, the crystallographer can jump ahead to later stages of the map-improvement process I am about to describe. If the molecular envelope cannot be discerned, then more data collection is required.

C. Improving the map

The crude molecular image seen in the F_0 map, which is obtained from the original indexed intensity data ($|F_{obs}|$) and the first phase estimates (α'_{calc}), serves now as a model of the desired structure. A crude electron density function is devised to describe the unit-cell contents as well as they can be observed in the first map. Then the function is modified to make it more realistic in the light of known properties of proteins and water in crystals. This process is called, depending on the exact details of procedure, *density modification*, *solvent leveling*, or *solvent flattening*.

The electron density function devised by density modification may be no more than a fixed, high value of $\rho(x,y,z)$ for all regions that appear to be within a protein molecule, and a fixed, low value of ρ for all surrounding areas of bulk solvent. One automated method first defines the molecular envelope by dividing the unit cell into a grid of regularly spaced points. At each point, the value of $\rho(x,y,z)$ in the F_0 map is evaluated. At each grid point, if ρ is negative, it is reassigned a value of zero; if ρ is positive, it is assigned a value equal to the average value of ρ within a defined distance of the gridpoint. This procedure smooths the map (eliminates many small, random fluctuations in density) and essentially divides the map into two types of regions: those of relatively high (protein) and relatively low (solvent) density. Next, the overall amplitude of the map is increased until the ratio of high density to low density agrees with the ratio of protein to solvent in the crystal, either assuming that the crystal is about half water, or using a value derived from the measured crystal density (Chapter 3, Section IV). This contrived function $\rho(x,y,z)$ is now used to compute structure factors, using Eq. (5.16). From this computation, we learn what the amplitudes and phases of all reflections would be if this new model were correct. We use the phases from this computation, which constitute a new set of α'_{hkl}s, along with the $|F_{obs}|$s derived from the original measured intensities, to calculate $\rho(x,y,z)$ again, using Eq. (7.3).

We do not throw out old phases immediately, but continue to weight each Fourier term with some measure of phase quality. In this manner, we continue to let the data speak for itself as much as possible, rather than allowing the current model to bias the results. If the new phase estimates are better, then the new $\rho(x,y,z)$ will be improved, and the electron-density map will be more detailed. The new map serves to define the molecular boundary more precisely, and the cycle is repeated. (Refer again to the block diagram in Fig. 7.1.) If we continue to use good judgment in incorporating new phases and new terms into Eq. (7.3), successive Fourier-series computations of $\rho(x,y,z)$ include more terms, and successive contour maps become clearer and more interpretable. In other words, the iterative process of incorporating phases from successively better and more complete models converges toward a structure that fits the native data better. The phase estimates "converge" in the sense that the output phases computed from the current model [Eq. (5.16)] agree better with the input phases that went into computation of the model [Eq. (7.3)].

As this process continues, and the model becomes more detailed, we begin to get estimates for the phases of structure factors at resolution beyond that of the heavy-atom derivatives. In a process called *phase extension*, we gradually increase the number of terms in the Fourier series of Eq. (7.3), adding terms that contain native intensities (as $|F_{obs}|$) at slightly higher resolution with phases from the current model. This must be done gradually and judiciously, so as not to let incorrect areas of the current model bias the calculations excessively. If the new phase estimates are good, the resulting map has slightly higher resolution, and structure factors computed from Eq. (5.16) give useful phase estimates at still higher resolution. In this manner, low-resolution phases are improved, and phase assignments are extended to higher resolution.

If phase extension seems like getting something from nothing, realize that by using general knowledge about protein and solvent density, we impose justifiable restrictions on the model, giving it realistic properties that are not visible in the map. In effect, we are using known crystal properties to increase the resolution of the model. Thus it is not surprising that the phases calculated from the modified model are good to higher resolution than those calculated from an electron-density function that little more than describes what can be seen in the map.

Another means of improving the map at this stage depends upon the presence of noncrystallographic symmetry elements in the unit cell. Recall that the intensity of reflections results from many molecules in identical orientations diffracting identically. In a sense, the diffraction pattern is the sum of diffraction patterns from all individual molecules. This is equivalent to taking a large number of weak, noisy signals (each the diffraction from one molecule) and adding them together to produce a strong signal. The noise in the individual signals, which might include the background intensity of the film or the weak signal of stray X rays, is random, and when many weak signals are added, this random noise cancels out.

In some cases, the strength of this signal can be increased further by averaging the signals from molecules that are identical but have different orientations in the unit cell, such that no two orientations of the crystal during data collection gives the same orientation of these molecules in the X-ray beam. These molecules may be related by symmetry elements that are not aligned with symmetry elements of the entire unit cell. Thus the diffractive contributions of these identical molecules are never added together. In such cases, the unit cell is said to exhibit *noncrystallographic symmetry*. By knowing the arrangement of molecules in the unit cell, that is, by knowing the location and type of noncrystallographic symmetry elements, the crystallographer can use a computer to simulate the movement of these sets of molecules into identical orientations and thus add their signals together. The result is improved signal-to-noise ratio and, in the end, a clearer image of the molecules. This method, called *symmetry averaging*, is spectacularly successful in systems with high symmetry, such as viruses. Many virus coat proteins are icosahedral, possessing two-, three-, and fivefold rotation axes. Often one or more two- and three-fold axes are noncrystallographic, and fivefold axes are always noncrystallographic, because no unit cell exhibits fivefold symmetry.

IV. The model becomes molecular

A. New phases from the molecular model

At some critical point in the iterative improvement of phases, the map becomes clear enough that we can trace the protein chain through it. In the worst circumstances at this stage, we may only be able to see some continuous tubes of density. Various aids may be used at this point to help the viewer trace the protein chain through the map. One is to *skeletonize* the map, which means to draw line segments along lines of maximum density. These so-called ridge lines show the viewer rough lines along which the molecular chain is likely to lie. They can help to locate both the main chain and the branch points of side chains.

In a clearer map, we may be able to recognize alpha helices, one of the densest features of a protein, or sheets of beta structure. Now we can construct a partial molecular model (as opposed to an electron-density model) of the protein, using computer graphics to build and manipulate a stick model of the known sequence within small sections of the map. This technique is called *map fitting*, and is discussed later. From the resulting model, which may harbor many errors and undefined regions, we again calculate structure factors, this time using Eq. (5.15), which treats each atom in the current model as an independent scatterer. In other words, we calculate new structure factors from our current, usually crude, molecular model rather than from an approximation of $\rho(x,y,z)$. Additional iterations may improve the map further, allowing more features to be constructed therein.

Here again, as in density modification, we are using known properties of proteins to improve the model beyond what we can actually see in the map. Thus we are in effect improving the resolution of the model by making it structurally realistic: giving it local electron densities corresponding to the light atoms we know are present and connecting atoms at bond lengths and angles that we know must be correct. So again, our successive models give us phases for reflections at higher and higher resolution. Electron-density maps computed from these phases and, as always, the original native amplitudes $|\mathbf{F}_{obs}|$ become more and more detailed.

B. Minimizing bias from the model

Conversion to a molecular model greatly increases the hazard of introducing excessive bias from the model into $\rho(x,y,z)$. At this point, bias can be decreased by one of several alternative Fourier computations of the electron-density map. As phases from the model begin to be the most reliable, they begin to dominate the Fourier series. In the extreme, the series would contain amplitudes purely from the intensity data and phases purely from the model. In order to compensate for the increased influence of model phases, and to continue letting the intensity data influence improvement of the model, the crystallographer calculates electron-density maps using various difference Fourier syntheses, in which the amplitude of each term is of the form $(|n|\mathbf{F}_{obs}| - |\mathbf{F}_{calc}||)$, which reduces overall model influence by subtracting the calculated structure-factor amplitudes ($|\mathbf{F}_{calc}|$) from some multiple of the observed amplitudes ($|\mathbf{F}_{obs}|$) within each Fourier term. For $n = 1$, the Fourier series is called an $\mathbf{F}_o - \mathbf{F}_c$ synthesis:

$$\rho(x,y,z) = \frac{1}{V}\sum_{h}\sum_{k}\sum_{l} \left||\mathbf{F}_o| - |\mathbf{F}_o|\right| e^{-2\pi i(hx+ky+lz-\alpha'_{calc})}.$$

(7.4)

A contour map of this Fourier series is called an $\mathbf{F}_o - \mathbf{F}_c$ map. How is this map interpreted? Depending on which of $\mathbf{F}_o$ or $\mathbf{F}_c$ is larger, Fourier terms can be either positive or negative. The resulting electron-density map contains both positive and negative "density." Positive density in a region of the map implies that the contribution of the observed intensities ($\mathbf{F}_o$s) to ρ are larger than the contribution of the model ($\mathbf{F}_c$s), and thus that the unit cell (represented by $\mathbf{F}_o$s) contains more electron density in this region than implied by the model (represented by $\mathbf{F}_c$s). In other words, the map is telling us that the model should be adjusted to increase the electron density in this region, by moving atoms toward the region. On the other hand, a region of negative density indicates that the model implies more electron density in the region than the unit cell actually contains. The region of negative density is telling us to move atoms away from this region. As an example, if an amino-acid side chain in the model is in the wrong conformation, the $\mathbf{F}_o - \mathbf{F}_c$ map may exhibit a negative peak coincident with the erroneous model side chain and a nearby positive peak signifying the correct position.

The $\mathbf{F}_o - \mathbf{F}_c$ map emphasizes errors in the current model, but it lacks the familiar appearance of the molecular surface found in an $\mathbf{F}_o$ map. In addition, if the model still contains many errors, the $\mathbf{F}_o - \mathbf{F}_c$ map is "noisy," full of small positive and negative peaks that are difficult to interpret. The $\mathbf{F}_o - \mathbf{F}_c$ map is most useful near the end of the structure determination, when most of the model errors have been eliminated. The $\mathbf{F}_o - \mathbf{F}_c$ map is a great aid in detecting subtle errors after most of the serious errors are corrected.

A more easily interpreted and intuitively satisfying difference map, but one that still allows undue influence by the model to be detected, is the $2\mathbf{F}_o - \mathbf{F}_c$ map, calculated as follows

$$\rho(x,y,z) = \frac{1}{V}\sum_h \sum_k \sum_l \left|2|\mathbf{F}_o| - |\mathbf{F}_o|\right| e^{-2\pi i (hx+ky+lz-\alpha'_{calc})}. \tag{7.5}$$

In this map, the model influence is reduced, but not as severely as with $\mathbf{F}_o - \mathbf{F}_c$. Unless the model contains extremely serious errors, this map is everywhere positive, and contours at carefully chosen electron densities resemble a molecular surface. With experience, the crystallographer can often see the bias of an incorrect area of the model superimposed upon the true signal of the correct structure as implied by the original intensity data. For instance, in a well-refined map (see Section V), backbone carbonyl oxygens are found under a distinct bulge in the backbone electron density. If a carbonyl oxygen in the model is pointing 180° away from the actual position in the molecule, the bulge in the map may be weaker than usual, or misshapen (sometimes cylindrical) and a weak bulge may be visible on the opposite side of the carbonyl carbon, at the true oxygen position. Correcting the oxygen orientation in the model, and then recalculating structure factors, results in loss of the weak, incorrect bulge in the map and intensification of the bulge in the correct location. (This may sound like a serious correction of the model, requiring the movement of many atoms, but the entire peptide bond can be flipped 180° around the backbone axis with only slight changes in the positions of neighboring atoms.)

Various other Fourier syntheses are used during these stages in order to improve the model. Some crystallographers prefer a $3\mathbf{F}_o - 2\mathbf{F}_c$ map, a compromise between $\mathbf{F}_o - \mathbf{F}_c$ and $2\mathbf{F}_o - \mathbf{F}_c$, for the final interpretation. In areas where the maps continue to be ambiguous, it is often helpful to examine the original MIR or molecular replacement maps for insight into how model building in this area might be started off on a different foot. Another measure is to eliminate the atoms in the questionable region and calculate structure factors from Eq. (5.15), so that the possible errors in the region contribute nothing to the phases, and hence do not bias the resulting map, which is called an *omit map*. (Another important type of difference Fourier synthesis, which is used to compare similar protein structures, is discussed in Chapter 8, Section III.C.)

C. Map fitting

Conversion to a molecular model is usually done piecemeal, as the map reveals recognizable structural features. This procedure, called *map fitting* or *model building*, entails interpreting the electron-density map by building a molecular model that fits realistically into the molecular surface implied by the map. Map fitting is done by interactive computer graphics. A computer program produces a realistic three-dimensional display of small sections of one or more electron-density maps, and allows the user to construct and manipulate molecular models to fit the map. As mentioned earlier, such programs can draw ridge lines to help the viewer trace the chain through areas of weak density.

As the model is built, the viewer sees the model within the map, as shown in Plate 2*b*. As the model is constructed or adjusted, the program stores current atom locations in the form of three-dimensional coordinates. The crystallographer, while building a model interactively on the computer screen, is actually building a list of atoms, each with a set of coordinates (x,y,z) to specify its location. Coordinates are automatically updated whenever the model is adjusted. This list of coordinates is the output file from the map-fitting program and the input file for calculation of new structure factors. When the model is correct and complete, this file becomes the means by which the model is shared with the community of scientists who study proteins (see Section VII).

In addition to routine commands for inserting or changing amino-acid residues, moving atoms and fragments, and changing conformations, map-fitting programs contain many sophisticated tools to aid the model builder. Fragments, treated as rigid assemblies of atoms, can be automatically fitted to the map by the method of least squares (see Section V.A). After manual adjustments of the model, which may result in unrealistic bond lengths and angles, portions of the model can be *regularized*, which entails automatic correction of bond lengths and angles with minimal movement of atoms. In effect, regularization looks for the most realistic configuration of the model that is very similar to its current configuration. Where small segments of the known sequence cannot be easily fitted to the map, some map-fitting programs can search fragment databases or the Protein Data Bank (see Section VII) for fragments having the same sequence, and then display these fragments so the user can see whether they fit the map.

Following is a somewhat idealized description of how map fitting may proceed, illustrated with views from a modern map-fitting program. The maps and models are from the structure determination of adipocyte lipid binding protein (ALBP), which I will discuss further in Chapter 8.

When the map has been improved to the point that molecular features are revealed, the crystallographer attempts to trace the protein through as much continuous density as possible. At this point, the quality of the map will vary from place to place, perhaps quite clear in the molecular interior, which is usually more ordered, and exhibiting broken density in some places, particularly at chain termini and surface loops. Because we know that amino-acid side chains branch regularly off alpha carbons in the main chain, we can estimate the positions of many alpha carbons. These atoms should lie near the center of the main-chain density next to bulges that represent side chains. In proteins, alpha carbons are 3.8–4.2 Å apart. This knowledge allows the crystallographer to construct an alpha-carbon model of the molecule (Plate 10) and to compute structure factors from this model.

Further improvement of the map with these phases may reveal side chains more clearly. Now the trick is to identify some specific side chains so that the known amino-acid sequence of the protein can be aligned with visible features in the map. As mentioned earlier, chain termini are often ill-defined, so we need a foothold for alignment of sequence with map where the map is sharp. Many times the key is a short stretch of sequence containing several bulky hydrophobic residues, like Trp, Phe, and Tyr. Because they are hydrophobic, they are likely to be in the interior where the map is clearer. Because they are bulky, their side-chain density is more likely to be identifiable. From such a foothold, the detailed model building can begin.

Regions that cannot be aligned with sequence are often built with polyalanine, reflecting our knowledge that all amino acids contain the same backbone atoms, and all but one, glycine, have at least a beta carbon (Plate 11). In this manner, we build as many atoms into the model as possible in the face of our ignorance about how to align the sequence with the map in certain areas.

In pleated sheets, we know that successive carbonyl oxygens point in opposite directions. One or two carbonyls whose orientations are clearly revealed by the map can allow sensible guesses as to the positions of others within the same sheet. As mentioned previously with respect to map fitting, we use knowledge of protein structure to infer more than the map shows us. If our inferences are correct, subsequent maps, computed with phases calculated from the model, will show enhanced evidence for the inferred features and will show additional features as well, leading to further improvement of the model. Poor inferences degrade the map, so where electron density conflicts with intuition, we follow the density as closely as possible.

With each successive map, new molecular features are added as they can be discerned, and errors in the model, such as side-chain conformations that no longer fit the electron density, are corrected. As the structure nears completion, the crystallographer may use $2F_o - F_c$ and $F_o - F_c$ maps simultaneously to track down the most subtle disagreements between the model and the data.

V. Structure refinement

A. Least-squares methods

Cycles of map calculation and model building, which are forms of real-space refinement of the model, are interspersed with computerized attempts to improve the agreement of the model with the original intensity data. (Everything goes back to those original reflection intensities, which give us our $|\mathbf{F}_{obs}|$ values!) Because these computations entail comparison of computed with observed structure factor amplitudes (reciprocal space), rather than examination of maps and models (real space), these methods are referred to as *reciprocal-space refinement.* Most commonly, this process is a massive version of least-squares fitting, the same procedure that freshman chemistry students employ to construct a straight line that fits a scatter graph of data.

In the simple least-squares method in two dimensions, the aim is to find a function $y = f(x)$ that fits a series of observations (x_1,y_1), (x_2,y_2),...(x_i,y_i), where each observation is a data point, a measured value of the independent variable x at some selected value y. (For example, y might be the temperature of a gas, and x might be its measured pressure.) The solution to the problem is a function $f(x)$ for which the sum of the squares of distances between the data points and the function itself is as small as possible. In other words, $f(x)$ is the function that minimizes D, the sum of the squared differences between observed (y_i) and calculated $[f(x_i)]$ values, as follows

$$D = \sum_i w_i [y_i - f(x_i)]^2. \qquad (7.6)$$

The differences are squared to make them all positive; otherwise, for a large number of random differences, D simply equals zero. The term w_i is an optional weighting factor that reflects the reliability of observation i, thus giving greater influence to the most reliable data. According to principles of statistics, w_i should be $1/(\sigma_i)^2$, where σ_i is the standard deviation computed from multiple measurements of the same data point (x_i,y_i).

In the simplest case, $f(x)$ is a straight line, for which the general equation is $f(x) = mx + b$, where m is the slope of the line and b is the intercept of the line on the $f(x)$-axis. Solving this problem entails finding the proper values of the parameters m and b. If we substitute $(mx_i + b)$ for each $f(x_i)$ in Eq. (7.6), take the partial derivative of the right-hand side with respect to m and set it equal to zero, and then take the partial derivative with respect to b and set it equal to zero, the result is a set of simultaneous equations in m and b. Because all the squared differences are to be minimized simultaneously, the number of equations equals the number of observations, and there must be at least two observations to fix values for the two parameters m and b. With just two observations (x_1, y_1) and (x_2, y_2), m and b are determined precisely, and $f(x)$ is the equation of the straight line between (x_1, y_1) and (x_2, y_2). If there are more than two observations, the problem is overdetermined and the values of m and b describe the straight line of best fit to all the observations. So the solution to this simple least-squares problem is a pair of parameters m and b for which the function $f(x) = mx + b$ minimizes D.

B. Crystallographic refinement

In the crystallographic case, the parameters we seek (analogous to m and b) are, for all atoms j, the positions (x_j, y_j, z_j) that best fit the observed structure-factor amplitudes. Because the positions of atoms in the current model can be used to calculate structure factors, and hence to compute the *expected* structure-factor amplitudes ($|\mathbf{F}_{calc}|$) for the current model, we want to find a set of atom positions that give $|\mathbf{F}_{calc}|$s, analogous to calculated values $f(x_i)$, that are as close as possible to the $|\mathbf{F}_{obs}|$s (analogous to observed values y_i). In least-squares terminology, we want to select atom positions that minimize the squares of differences between corresponding $|\mathbf{F}_{calc}|$s and $|\mathbf{F}_{obs}|$s. We define the difference between the observed amplitude $|\mathbf{F}_{obs}|$ and the measured amplitude $|\mathbf{F}_{calc}|$ for reflection hkl as $(|\mathbf{F}_o| - |\mathbf{F}_c|)_{hkl}$, and we seek to minimize the function Φ, where

$$\Phi = \sum_{hkl} w_{hkl} \left(|\mathbf{F}_o| - |\mathbf{F}_c| \right)^2_{hkl}. \qquad (7.7)$$

In words, the function Φ is the sum of the squares of differences between observed and calculated amplitudes. The sum is taken over all reflections hkl currently in use. Each difference is weighted by the term w_{hkl}, a number that depends on the reliability of the corresponding measured intensity. As in the simple example, according to principles of statistics, the weight should be $1/(\sigma_{hkl})^2$, where σ is the standard deviation from multiple measurements of $|\mathbf{F}_{obs}|$. Because the data do not usually contain enough measurements of each reflection to determine its standard deviation, other weighting schemes have been devised. Starting from a reasonable model, the least-squares refinement method succeeds about equally well with a variety of weighting systems, so I will not discuss them further.

C. Additional refinement parameters

We seek a set of parameters that minimize the function Φ. These parameters include the atom positions, of course, because the atom positions in the model determine each $\mathbf{F}_{calc}$. But other parameters are included as well. One is the *temperature factor B_j,* or *B*-factor, of each atom *j*, a measure of how much the atom oscillates around the position specified in the model. Atoms at side-chain termini are expected to exhibit more freedom of movement than main-chain atoms, and this movement amounts to spreading each atom over a small region of space. Diffraction is affected by this variation in atomic position, so it is realistic to assign a temperature factor to each atom and include the factor among parameters to vary in minimizing Φ. From the temperature factors computed during refinement, we learn which atoms in the molecule have the most freedom of movement, and we gain some insight into the dynamics of our largely static model. In addition, adding the effects of motion to our model makes it more realistic and hence more likely to fit the data precisely.

Another parameter included in refinement is the *occupancy n_j* of each atom *j*, a measure of the fraction of molecules in which atom *j* actually occupies the position specified in the model. If all molecules in the crystal are precisely identical, then occupancies for all atoms are 1.00. Occupancy is included among refinement parameters because occasionally two or more distinct conformations are observed for a small region like a surface side chain. The model might refine better if atoms in this region are assigned occupancies equal to the fraction of side chains in each conformation. For example, if the two conformations occur with equal frequency, then atoms involved receive occupancies of 0.5 in each of their two possible positions. By including occupancies among the refinement parameters, we obtain estimates of the frequency of alternative conformations, giving some additional information about the dynamics of the protein molecule.

The factor $|\mathbf{F}_c|$ in Eq. (7.7) can be expanded to show all the parameters included in refinement, as follows:

$$\mathbf{F}_c = G \cdot \sum_j n_j f_j e^{\,2\pi i(hx_j+ky_j+lz_j)} \cdot e^{-B_j[\sin\theta)/\lambda]^2}. \tag{7.8}$$

Although this equation is rather forbidding, it is actually a familiar equation (5.15) with the new parameters included. Equation (7.8) says that structure factor $\mathbf{F}_{hkl}$ can be calculated ($\mathbf{F}_c$) as a Fourier series containing one term for each atom *j* in the current model. *G* is an overall scale factor to put all $\mathbf{F}_c$s on a convenient numerical scale. In the *j*th term, which describes the diffractive contribution of atom *j* to this particular structure factor, n_j is the occupancy of atom *j*; f_j is its scattering factor, just as in Eq. (5.16); x_j, y_j, and z_j are its coordinates; and B_j is its temperature factor. The first exponential term is the familiar Fourier description of a simple three-dimensional wave with frequencies *h, k,* and *l* in the directions *x, y,* and *z*. The second exponential shows that the effect of B_j on the structure factor depends on the angle of the reflection $[(\sin\theta)/\lambda]$.

D. *Local minima and radius of convergence*

As you can imagine, finding parameters (atomic coordinates, occupancies, and temperature factors for all atoms in the model) to minimize the differences between all the observed and calculated structure factors is a massive computing task. As in the simple example, one way to solve this problem is to differentiate Φ with respect to all the parameters, which gives simultaneous equations with the parameters as unknowns. The number of equations equals the number of observations, in this case the number of measured reflection intensities in the native data set. The parameters are overdetermined only if the number of measured reflections is greater than the number of parameters to be obtained. The complexity of the equations rules out analytical solutions and requires iterative (successive-approximations) methods that we hope will converge from the starting parameters of our current model to a set of new parameters corresponding to a minimum in Φ. It has been proved that the atom positions that minimize Φ are the same as those found from Eq. (7.3), the Fourier-series description of electron density. So real-space and reciprocal-space methods converge to the same solution.

The complicated function Φ undoubtedly exhibits many *local minima*, corresponding to variations in model conformation that minimize Φ with respect to other quite similar ("neighboring") conformations. A least-squares procedure will find the minimum that is nearest the starting point, so it is important that the starting model parameters be near the *global minimum*, the one conformation that gives best agreement with the native structure factors. Otherwise the refinement will converge into an incorrect local minimum from which it cannot extract itself. The greatest distance from the global minimum from which refinement will converge properly is called the *radius of convergence*. The theoretically derived radius is $d_{min}/4$, where d_{min} is the lattice-plane spacing corresponding to the reflection of highest resolution used in the refinement. Inclusion of data from higher resolution, while potentially giving more information, decreases the radius of convergence, so the model must be ever closer to its global minimum as more data are included in refinement. There are a number of approaches to increasing the radius of convergence, and thus increasing the probability of finding the global minimum.

These approaches take the form of additional constraints and restraints on the model during refinement computations. A *constraint* is a fixed value for a certain parameter. For example, in early stages of refinement, we might constrain all occupancies to a value of 1.0. A *restraint* is a subsidiary condition imposed upon the parameters, such as the condition that all bond lengths and bond angles be within a specified range of values. The function Φ, with additional restraints on bond lengths and angles, is as follows:

$$\Phi = \sum_{hkl} w_{hkl} \left(|\mathbf{F_o}| - |\mathbf{F_c}| \right)^2_{hkl}.$$

$$+ \sum_{i}^{bonds} w_i \left(d_i^{ideal} - d_i^{model} \right)^2 \tag{7.9}$$

$$+ \sum_{j}^{angles} w_j \left(\phi_j^{ideal} - \phi_j^{model} \right)^2,$$

where d_i is the length of bond i, and ϕ_j is the bond angle at location j. Ideal values are average values for bond lengths and angles in small organic molecules, and model values are taken from the current model. In minimizing this more complicated Φ, we are seeking atom positions, temperature factors, and occupancies that simultaneously minimize differences between (1) observed and calculated reflection amplitudes, (2) model bond lengths and ideal bond lengths, and (3) model bond angles and ideal bond angles. In effect, the restraints penalize adjustments to parameters if the adjustments make the model less realistic.

E. Molecular energy and motion in refinement

Crystallographers can take advantage of the prodigious power of today's computers to include knowledge of molecular energy and molecular motion in the refinement. In *energy refinement*, least-squares restraints are placed upon the overall energy of the model, including bond, angle, and conformational energies and the energies of noncovalent interactions such as hydrogen bonds. Adding these restraints is an attempt to find the structure of lowest energy in the neighborhood of the current model. In effect, these restraints penalize adjustments to parameters if the adjustments increase the calculated energy of the model.

Another form of refinement employs *molecular dynamics*, which is an attempt to simulate the movement of molecules by solving Newton's laws of motion for atoms moving within force fields that represent the effects of covalent and noncovalent bonding. Molecular dynamics can be turned into a tool for crystallographic refinement by including an energy term that is related to the difference between the measured reflection intensities and the intensities calculated from the model. In effect, this approach treats the model as if its energy decreases as its fit to the native crystallographic data improves. In refinement by *simulated annealing*, the model is allowed to move as if at high temperature, in hopes of lifting it out of local energy minima. Then the model is cooled slowly to find its preferred conformation at the temperature of diffraction data collection. All the while, the computer is searching for the conformation of lowest energy, with the assigned energy partially dependent upon agreement with diffraction data. In some cases, the radius of convergence is greatly increased by this process, a form of molecular dynamics refinement.

VI. Convergence to a final structure

A. Producing the final map and model

In the last stages of structure determination, the crystallographer alternates computed, reciprocal-space refinement with map fitting, or real-space refinement. In general, constraints and restraints are lifted as refinement proceeds, so that agreement with the original reflection intensities is gradually given highest priority. When ordered water becomes discernible in the map, water molecules are added to the model, and occupancies are no longer constrained, to reflect the fact that a particular water site may be occupied in only a fraction of unit cells. Early in refinement, all temperature factors are assigned a starting value. Later, the value is held the same for all atoms or for groups of similar atoms (like all backbone atoms as one group and all side-chain atoms as a separate group), but the overall value is not constrained. Finally, individual atomic temperature factors are allowed to refine independently. Early in refinement, the whole model is held rigid to refine its position in the unit cell. Then blocks of the model are held rigid while their positions refine with respect to each other. In the end, individual atoms are freed to refine with only stereochemical restraints. This gradual release of the model to refine against the original data is an attempt to prevent it from getting stuck in local minima. Choosing when to relax specific constraints and restraints is perhaps more art than science.

Near the end of refinement, the $F_o - F_c$ map becomes rather empty except in problem areas. Map fitting becomes a matter of searching for and correcting errors in the model, which amounts to extricating the model from local minima in the reciprocal-space refinement. Wherever model atoms lie outside $2F_o - F_c$ contours, the $F_o - F_c$ map will often show the atoms within negative contours, with nearby positive contours pointing to correct locations for these atoms. Many crystalline proteins possess disordered regions, where the maps do not clear up and become unambiguously interpretable. Such regions of structural uncertainty are mentioned in published papers on the structure and in the header information of Protein Data Bank files (see Section VII).

At the end of successful refinement, the $2F_o - F_c$ map almost looks like a space-filling model of the protein. (Refer to Plate 2b, which is the final model built into the same region shown in Plates 10 and 11.) The backbone electron density is continuous, and peptide carbonyl oxygens are clearly marked by bulges in the backbone density. Side-chain density, especially in the interior, is sharp and fits the model snugly. Branched side chains, like those of valine, exhibit distinct lobes of density representing the two branches. Rings of histidine, phenylalanine, tyrosine, and tryptophan are flat, and in models of the highest resolution, aromatic rings show a clear depression or hole in the density at their centers. Looking at the final model in the final map, you can easily underestimate the difficulty of interpreting the early maps, in which backbone density is frequently weak and broken, and side chains are missing or shapeless.

You can get a rough idea of how refinement gradually reveals features of the molecule by comparing electron-density maps at low, medium, and high resolution, as in Plate 12. Each photo in this set shows a section of the final ALBP model in a map calculated with the final phases, but with $|\mathbf{F}_{obs}|$s limited to specified resolution. In (a), only $|\mathbf{F}_{obs}|$s of reflections at resolution 6 Å or greater are used. With this limit on the data (which amounts to including in the $2\mathbf{F}_o - \mathbf{F}_c$ Fourier series only those reflections whose indices hkl correspond to sets of planes with spacing d_{hkl} of 6 Å or greater), the map of this pleated-sheet region of the protein is no more than a featureless sandwich of electron density. As we extend the Fourier series to include reflections out to 4.5 Å, the map (b) shows distinct, but not always continuous, tubes of density for each chain. Extending the resolution to 3.0 Å, we see density that defines the final model reasonably well, including bulges for carbonyl oxygens (red) and for side chains. Finally, at 1.6 Å, the map fits the model like a glove, zigzagging precisely in unison with the backbone of the model and showing well-defined lobes for individual side-chain atoms.

Look again at the block diagram of Fig. 7.1, which gives an overview of structure determination. Now I can be more specific about the criteria for error removal or *filtering*, which is shown in the diagram as horizontal dashed lines in real and reciprocal space. Real-space filtering of the *map* entails removing noise or adding density information, as in solvent flattening. Reciprocal-space filtering of *phases* entails using only the strongest reflections (for which phases are more accurate) to compute the early maps, and using figures of merit and phase probabilities to select the most reliable phases at each stage. The molecular *model* can be filtered in either real or reciprocal space. Errors are removed in real space by improving the fit of model to map, and by allowing only realistic bond lengths and angles when adjusting the model (regularization). Here the criteria are structural parameters and congruence to the map (real space). Model errors are removed in reciprocal space (curved arrow in center) by least-squares refinement, which entails adjusting atom positions in order to bring calculated intensities into agreement with measured intensities. Here the criteria are comparative structure-factor amplitudes (reciprocal space). Using the

Fourier transform, the crystallographer moves back and forth between real and reciprocal space to nurse the model into congruence with the data.

B. Guides to convergence

Judging convergence and assessing model quality are overlapping tasks. I will discuss criteria of convergence here. In Chapter 8, I will discuss some of the criteria further, particularly as they relate to the quality and usefulness of the final model.

The progress of iterative real- and reciprocal-space refinement is monitored by comparing the measured structure-factor amplitudes $|\mathbf{F}_{obs}|$ (which are proportional to $(I_{obs})^{1/2}$) with amplitudes $|\mathbf{F}_{calc}|$ from the current model. In calculating the new phases at each stage, we learn what intensities our current model, if correct, would yield. As we converge to the correct structure, the measured $\mathbf{F}$s and the calculated $\mathbf{F}$s should also converge. The most widely used measure of convergence is the *residual index*, or R-factor (Chapter 6, Section V.E).

$$R = \frac{\sum \left| |\mathbf{F}_{obs}| - |\mathbf{F}_{calc}| \right|}{\sum |\mathbf{F}_{obs}|}. \qquad (7.10)$$

In this expression, each $|\mathbf{F}_{obs}|$ is derived from a measured reflection intensity and each $|\mathbf{F}_{calc}|$ is the amplitude of the corresponding structure factor calculated from the current model. Values of R range from zero, for perfect agreement of calculated and observed intensities, to about 0.6, the R-factor obtained when a set of measured amplitudes is compared with a set of random amplitudes. An R-factor greater than 0.5 implies that agreement between observed and calculated intensities is very poor, and many models with $R = 0.5$ or greater will not respond to attempts at improvement unless more data are available. An early model with R near 0.4 is promising and is likely to improve with the various refinement methods I have presented. A desirable target R-factor for a protein model refined with data to 2.5 Å is 0.2. Very rarely, small, well-ordered proteins may refine to $R = 0.1$, whereas small organic molecules commonly refine to better than $R = 0.05$.

A more demanding and revealing criterion of model quality and of improvements during refinement is the *free R-factor,* R_{free}. R_{free} is computed with a small set of randomly chosen intensities, the "test set," which are set aside from the beginning and not used during refinement. They are used only in the *cross-validation* or quality control process of assessing the agreement between calculated (from the model) and observed data. At any stage in refinement, R_{free} measures how well the current atomic model predicts a subset of the measured intensities that were not included in the refinement, whereas R measures how well the current model predicts the entire data set that produced the model. Many crystallographers believe that R_{free} gives a better and less-biased measure of overall model. In many test calculations, R_{free} correlates very well with phase accuracy of the atomic model. In general, during intermediate stages of refinement, R_{free} values are higher than R, but in the final stages, the two often become more similar. Because incompleteness of data can make structure determination more difficult (and perhaps because the lower values of R are somewhat seductive in stages where encouragement is welcome), some crystallographers at first resisted using R_{free}. But many now use both Rs to guide them in refinement, looking for refinement procedures that improve both Rs, and proceeding with great caution when the two criteria appear to be in conflict.

In addition to monitoring R-factors as indicators of convergence, the crystallographer monitors various structural parameters that indicate whether the model is chemically, stereochemically, and conformationally reasonable. In a *chemically* reasonable model, the bond lengths and bond angles fall near the expected values for simple organic molecules. The usual criteria applied are the root-mean-square (rms) deviations of all the model's bond lengths and angles from an accepted set of values. A well-refined model exhibits rms deviations of no more than 0.02 Å for bond lengths and 4° for bond angles.

A *stereochemically* reasonable model has no inverted centers of chirality (for instance, no D-amino acids). A *conformationally* reasonable model meets several criteria (1) Peptide bonds are nearly planar, and nonproline peptide bonds are *trans*-, except where obvious local conformational constraints produce an occasional *cis*-proline. (2) The backbone conformational angles Φ and Ψ fall in allowed ranges, as judged from Ramachandran plots of these angles (see Chapter 8); and finally, (3) torsional angles at single bonds in side chains lie within a few degrees of stable, staggered conformations. During the progress of refinement, all of these structural parameters should continually improve.

VII. Sharing the model

An intensely interested audience awaits the crystallographer's final molecular model. This audience includes researchers studying the same molecule by other methods, such as spectroscopy or kinetics, or studying metabolic pathways or diseases in which the molecule is involved. The model may serve as a basis for understanding the properties of the protein and its behavior in biological systems. It may also serve as a guide to the design of inhibitors or to engineering efforts to modify its function by methods of molecular biology.

Most crystallographers appear to believe that it is part and parcel of their work to make molecular structures available to the larger community of scientists. This belief is reflected in the policies of many journals and funding organizations that require public availability of the structure as a condition of publication or financial support. But in the desire to make sure that they do not fail to capitalize on commercially important leads provided by new models, crystallographers sometimes delay making models available. Research support from industry sometimes carries the stipulation that the full atomic details of new models be withheld long enough to allow researchers to explore and perhaps patent ideas of potential commercial value. Some journals now allow such delays, agreeing to publish announcements and discussions of new models if public access within a reasonable period is assured. This assurance can be enforced by requiring, as a condition of publication, that the researcher submit the model to the Protein Data Bank (discussed later in this section) immediately, but with the stipulation that public access is denied for a time, usually no more than one year.

Crystallographers share the fruits of their work in the form of lists of atomic coordinates, which can be used to display and study the molecule with molecular graphics programs (Chapter 11). Less commonly, because fewer people have the resources to use them, crystallographers share the final structure factors, from which electron-density maps can be computed. The audience for structure factors includes other crystallographers developing new techniques of data handling, refinement, or map interpretation.

Upon request, many authors of published crystallographic structures provide coordinate lists or structure factors by computer mail directly to interested parties. But the great majority of structures are available through the Protein Data Bank (PDB).[1] Crystallographers can satisfy publication and funding requirements for availability of their structures by depositing coordinates with this data bank.

The Protein Data Bank checks deposited files carefully for errors and inconsistencies and then makes them available free over the World Wide Web, in a standard text format. The PDB structure files, which are called *atomic*

[1]The Protein Data Bank is described fully in F. C. Bernstein, T. F. Koetzle, G. J. B. Williams, E. F. Meyer, Jr., M. D. Brice, J. R. Rodgers, O. Kennard, T. Shimanouchi, and M. Tasume, The Protein Data Bank: A computer-based archival file for macromolecular structures, *J. Mol. Biol.* 112, 535–542, 1977, and E. E. Abola, F. C. Bernstein, S. H. Bryant, T. F. Koetzle, and J. Weng, "Protein Data Bank," in F. H. Allen, F. Bergerhoff, and R. Sievers, eds., *Crystallographic Database—Information Content, Software Systems, Scientific Applications,* Data Commission of the International Union of Crystallography, Bonn-Cambridge-Chester, 1987, pp. 107–132.

coordinate entries, can be read within editor or word-processor programs. Almost all molecular graphics programs read PDB files directly or use them to produce their own files in binary form for rapid access during display. In addition to the coordinate list, a PDB file contains a *header* or opening section with information about published papers on the protein, details of experimental work that produced the structure, and other useful information.

Here is a brief description of PDB file contents. The line types, given in capital letters, are printed at the left of each line in the file. Some of these line types do not apply to all models and may be missing.

The contents of the file, in order of appearance, are

- HEADER and TITLE lines, containing the file name, date, and a brief title.
- COMPND lines, containing the name of the protein.
- SOURCE lines, giving the organism from which the protein was obtained.
- KEYWDS lines, giving keywords that would guide a search to this file.
- AUTHOR lines, listing the persons who placed this data in the Protein Data Bank.
- REVDAT lines, listing all revision dates for data on this protein.
- JRNL lines, giving the journal reference to the lead article about this model.
- REMARK lines, containing (1) references to journal articles about the structure of this protein and (2) general information about the contents of this file, including many specifics about refinement methods and final criteria of model quality (see Chapter 8).
- SEQRES lines, giving the amino-acid sequence of the protein, with amino acids specified by three-letter abbreviation.
- HET and FORMUL lines, listing the cofactors, prosthetic groups, or other nonprotein substances present in the structure. On-line versions of PDB files often contain links to more information about HET groups, including links to graphics displays of their structures (see Chapter 11).
- HELIX, SHEET, TURN, CISPEP, and SITE lines, listing the elements of secondary structure in the protein, residues involved in *cis*-peptide bonds (almost always involving proline as the second residue), and residues in the active site of the protein.
- CRYST lines, giving the unit cell dimensions and space group.
- ORIG and SCALE lines, containing instructions for computing the positions of symmetry-related molecules in the unit cell.
- ATOM lines, containing the atomic coordinates of all protein atoms, plus their structure factors and occupancies. Atoms are listed in the order given in the paragraph following this list.

- TER lines, among the ATOM lines, specifying the termini of distinct chains in the model.
- HETATM lines, which contain the same information as ATOM lines for any *nonprotein* molecules (cofactors, prosthetic groups, and solvent molecules, collectively called *heteromers*) included in the structure and listed in HET and FORMUL lines above.
- CONECT lines, which list bonds between nonprotein atoms in the file.
- MASTER and END lines, which mark the end of the file.

After the header comes a list of model atoms in standard order. Atoms in the PDB file are named and listed according to a standard format in an all-English version of the Greek-letter conventions used by organic chemists. For each amino acid, beginning at the N terminus, the backbone atoms are listed in the order alpha nitrogen N, alpha carbon CA, carbonyl carbon C, and carbonyl oxygen O, followed by the side chain atoms, beta carbon CB, gamma carbon CG, and so forth. In branched side chains (or rings), atoms in the two branches are numbered 1 and 2 after the proper Greek letter. For example, the atoms of aspartic acid, in the order of PDB format, are N, CA, C, O, CB, CG, OE1, and OE2. The terminal atoms of the side chain are followed in the file by atom N of the next residue. There are no markers in the file to tell where one residue begins and another ends; each N marks the beginning of the next residue.

In this form, as a PDB atomic coordinate entry, a crystallographic structure becomes a matter of public record. The final model of the molecule can then fall before the eyes of anyone equipped with a computer and an appropriate molecular display program. It is natural for the consumer of these files, as well as for anyone who sees published structures in journals or textbooks, to think of the molecule as something someone has seen more or less directly. Having read this far, you know that our crystallographic vision is quite indirect. But you probably still have little intuition about possible limits to the model's usefulness. For instance, just how precise are the relative locations of atoms? How much does molecular motion alter atomic positions? For that matter, how well does the model fit the original diffraction data from which it was extracted? These and other questions are the subject of Chapter 8, in which I will start you off toward becoming a discriminating consumer of the crystallographic product. This entails understanding several criteria of model quality and being able to extract these criteria from published accounts of crystallographic structure determination.

8   A User's Guide to Crystallographic Models

I. Introduction

Most biochemists will never determine a protein structure by X-ray crystallography. But many will at some time use a crystallographic model in research or teaching. In research, study of molecular models by computer graphics is an indispensable tool in formulating mechanisms of protein action (for instance, binding or catalysis), searching for modes of interaction between molecules, choosing sites to modify by chemical methods or site-specific mutagenesis, and designing inhibitors of proteins involved in disease. Because protein chemists would like to learn the rules of protein folding, every new model is a potential test for proposed theories of folding, as well as for schemes for predicting conformation from amino-acid sequence. Every new model for which homologous sequences are known is a potential scaffold on which to build homology models (see Chapter 11). In education, modern texts in biology and chemistry are effectively and dramatically illustrated with graphics images, often as stereo pairs. Projection monitors allow instructors to show "real-time" graphics displays in the classroom, giving students vivid, animated, three-dimensional views of complex molecules. Many introductory biochemistry courses now include the study of macromolecules using molecular graphics programs on personal computers. Free or inexpensive, yet very powerful, graphics programs for personal computers, combined with easy access to the Protein Data Bank over the World Wide Web, now make it possible for anyone to study any available molecular model.

In all of these applications, there is a tendency to treat the model as a physical entity, as a real object seen or filmed. How much confidence in the crystallographic model is justified? For instance, how precisely does crystallography establish the positions of atoms in the molecule? Are all of the atomic positions equally well established? How does one rule out the possibility that crystallizing the protein alters it in some significant way? The model is a static image of a dynamic molecule, a springy system of atoms that breathes with characteristic vibrations, and tumbles dizzily through solution, as it executes its function. Does crystallography give us any insight into these motions? Are parts of the molecule more flexible than others? Are major movements of structural elements essential to the molecule's action? How does the user decide whether proposed motions of the molecule are reasonable?

In this chapter, I will discuss the strengths and limitations of molecular models obtained by X-ray diffraction. My aim is to help you to use crystallographic models wisely and appropriately, and realize just what is known, and what is unknown, about a molecule that has yielded up some of its secrets to crystallographic analysis. To demonstrate how you can draw these conclusions for yourself with regard to a particular molecule of interest, I will conclude this chapter by discussing a recent structure determination, as it appeared in a biochemical journal. Here my goals are (1) to help you learn to extract criteria of model quality from published structural reports and (2) to review some basic concepts of protein crystallography.

II. Judging the quality and usefulness of the refined model

A. Structural parameters

As discussed in Chapter 7, Section VI.B, crystallographers monitor an R-factor as an indicator of convergence to a final, refined model, with a general target of 0.20 or lower for proteins, and adequate additional cycles of refinement to confirm that R is not still declining. In addition, various constraints and restraints are relaxed during refinement, and after these restricted values are allowed to refine freely, they should remain in, or converge to, reasonable values. Among these are the root-mean-square (rms) deviations of the model's bond lengths, angles, and conformational angles from an accepted set of values based upon the geometry of small organic molecules. A refined model should exhibit rms deviations of no more than 0.02 Å for bond lengths and 4° for bond angles. These values are routinely calculated during refinement to be sure that all is going well.

In effect, protein structure determination is a search for the conformation of a molecule whose chemical composition is known. For this reason, conformational angles about single bonds are not constrained during refinement, and they should settle into reasonable values. Spectroscopic evidence abundantly implies that peptide bonds are planar, and some refinements constrain peptide geometry. If unconstrained, peptide bonds should settle down to within one to two degrees of planar.

The other backbone conformational angles are Φ, along the N–Cα bond and Ψ, along the Cα–C bond, as shown in Fig. 8.1. In this figure, Φ is the torsional angle of the N-Cα bond, defined by the atoms C–N– Cα–C (C is the carbonyl carbon) and Ψ is the torsional angle of the Cα–C bond, defined by the atoms N– Cα–C–N. In the figure, $\Phi = \Psi = 180°$.

Model studies show that, for each amino acid, the pair of angles Φ and Ψ is greatly restricted by steric repulsion. The allowed pairs of values are depicted on a Ramachandran diagram (Fig. 8.2). A point (Φ,Ψ) on the diagram represents the conformational angles Φ and Ψ on either side of the alpha carbon of one residue.

Irregular polygons enclose backbone conformational angles that do not give steric repulsion (inner polygons) or give only modest repulsion (outer polygons). Location of the letters α and β correspond to conformational angles of residues in α helix and β pleated sheet.

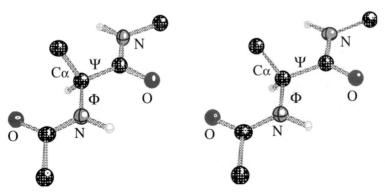

Figure 8.1 Backbone conformational angles in proteins (stereo).

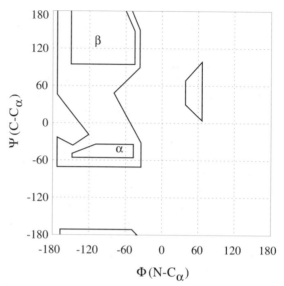

Figure 8.2   Ramachandran diagram for nonglycine amino-acid residues in proteins. Angles Φ and Ψ are as defined in Fig. 8.1.

During the final stages of map fitting and crystallographic refinement, Ramachandran diagrams are a great aid in finding conformationally unrealistic regions of the model. Crystallographic software packages and map-fitting programs usually contain a routine for computing Φ and Ψ for each residue from the current coordinate list, as well as for generating the Ramachandran diagram and plotting a symbol or residue number at the position (Φ,Ψ). Refinement papers often include the diagram, with an explanation of any residues that lie in high-energy ("forbidden") areas. For an example, see Fig. 8.6 in Section III.C. Glycines, because they lack a side chain, usually account for most of the residues that lie outside allowed regions. If nonglycine residues exhibit forbidden conformational angles, there should be some explanation in terms of structural constraints that overcome the energetic cost of an unusual backbone conformation.

The conformations of amino-acid side chains are unrestrained during refinement. In well-refined models, side-chain single bonds end up in staggered conformations.

B. Resolution and precision of atomic positions

In microscopy, the phrase "resolution of 2 Å" implies that we can resolve objects that are 2 Å apart. If this phrase had the same meaning for a crystallographic model of a protein, in which bond distances average about 1.5 Å, we would be unable to distinguish or resolve adjacent atoms in a 2-Å map. Actually, for a protein refined at 2-Å resolution to an R-factor near 0.2, the situation is much better than the resolution statement seems to imply.

In X-ray crystallography, "2-Å model" means that analysis included reflections out to a distance in the reciprocal lattice of 1/(2 Å) from the center of the diffraction pattern. This means that the model takes into account diffraction from sets of equivalent, parallel planes spaced as closely as 2 Å in the unit cell. (Presumably, data farther out than the stated resolution was unobtainable or was too weak to be reliable.) Although the final 2-Å map, viewed as an empty contour surface, may indeed not allow us to discern adjacent atoms, structural constraints on the model greatly increase the precision of atom positions. The main constraint is that we know we can fit the map with *groups* of atoms—amino-acid residues—having known connectivities, bond lengths, bond angles, and stereochemistry.

More than the resolution, we would like to know the precision with which atoms in the model have been located. For years, crystallographers used the Luzzati plot (Fig. 8.3) to estimate the precision of atom locations in a refined crystallographic model. At best, this is an estimate of the upper limit of error in atomic coordinates. The numbers to the right of each smooth curve on the Luzzati plot are theoretical estimates of the average uncertainty in the positions of atoms in the refined model (more precisely, the rms errors in atom positions). The average uncertainty has been shown to depend upon R-factors derived from the final model in various resolution ranges. To prepare data for a Luzzati plot, we separate the intensity data into groups of reflections in

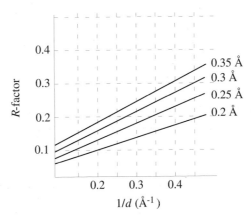

Figure 8.3   Luzzati diagram.

narrow ranges of $1/d$ (where d is the spacing of real lattice planes). Then we plot each R-factor (vertical axis) versus the midpoint value of $1/d$ for that group of reflections (horizontal axis). For example, we calculate R using only reflections corresponding to the range $1/d = 0.395–0.405$ (reflections in the 2.53- to 2.47-Å range) and plot this R-factor versus $1/d = 0.400$/Å, the midpoint value for this group. We repeat this process for the range $1/d \doteq 0.385–0.395$, and so forth. As the theoretical curves indicate, the R-factor typically increases for lower-resolution data (higher values of $1/d$). The resulting curve should roughly fit one of the theoretical curves on the Luzzati plot. From the theoretical curve closest to the experimental R-factor curve, we learn the average uncertainty in the atom positions of the final model. It has been claimed that Luzzati plots with R_{free} (Chapter 7, Section VI.B) give even better estimates of uncertainty in coordinates.

Publications, especially older ones, of refined structures may include a Luzzati plot, allowing the reader to assess very roughly the average uncertainty of atom positions in the model. Alternatively, they may simply report the uncertainty as "determined by the method of Luzzati." For highly refined models, rms errors as low as 0.15 Å are sometimes attained. In Fig. 8.5 a, Section III.C, the jagged curve represents the data for the refined model of the protein ALBP. The position of the curve on the Luzzati plot indicates that rms error for this model is about 0.34 Å, about one-fifth the length of a carbon-carbon bond.

In crystallography, unlike microscopy, the term *resolution* simply refers to the amount of data ultimately phased and used in the structure determination. In contrast, the precision of atom positions depends in part upon the resolution limits of the data, but also depends critically upon the quality of the data, as reflected by the R-factor. Good data can yield atom positions that are precise to within one-fifth to one-tenth of the stated resolution.

C. Vibration and disorder

Notice, however, that the preceding analysis gives only an upper limit and an average, or rms value, of position errors, and further, that the errors result from the limits of accuracy in the data. There are also two important physical (as opposed to statistical) reasons for uncertainty in atom positions: thermal motion and disorder. *Thermal motion* refers to vibration of an atom about its rest position. *Disorder* refers to atoms or groups of atoms that do not occupy the same position in every unit cell, in every asymmetric unit, or in every molecule within an asymmetric unit. The temperature factor B_j obtained during refinement reflects both the thermal motion and the disorder of atom j, making it difficult to sort out these two sources of uncertainty.

Occupancies n_j for atoms of the protein (but not necessarily its ligands, which may be present at lower occupancies) are usually constrained at 1.0 early in refinement, and in many refinements are never released, so that both thermal motion and disorder show their effects upon the final B values. In some cases, after refinement converges, a few B values fall far outside the average range for the model. This is sometimes an indication of disorder. Careful examination of $2F_o - F_c$ and $F_o - F_c$ maps may give evidence for more than one conformation in such a troublesome region. If so, inclusion of multiple conformations followed by refinement of their occupancies may improve the R-factor and the map, revealing the nature of the disorder more clearly.

If B_j were purely a measure of thermal motion at atom j (and assuming that occupancies are correct), then in the simplest case of purely harmonic thermal motion of equal magnitude in all directions (called *isotropic* vibration), B_j is related to the magnitude of vibration as follows:

$$B_j = 8\pi^2 \{u_j^2\} = 79 \{u_j^2\}, \tag{8.1}$$

where $\{u_j^2\}$ is the mean-square displacement of the atom from its rest position. Thus if the measured B_j is 79 Å^2, the total mean-square displacement of atom j due to vibration is 1.0 Å^2, and the rms displacement is the square root of $\{u_j^2\}$, or 1.0 Å. The B values of 20 and 5 Å^2 correspond to rms displacements of 0.5 and 0.25 Å. But the B values obtained for most proteins are too large to be seen as reflecting purely thermal motion and must certainly reflect disorder as well.

With small molecules, it is usually possible to obtain anisotropic temperature factors during refinement, giving a picture of the preferred directions of vibration for each atom. But a description of anisotropic vibration requires six parameters per atom, vastly increasing the computational task. In many cases, the total number of parameters sought, including three atomic coordinates, one occupancy, and six thermal parameters per atom, approaches or exceeds the number of measured reflections. As mentioned earlier, for refinement to succeed, observations (measured reflections and constraints such as bond lengths) must outnumber the desired parameters, so that least-squares solutions are adequately overdetermined. For this reason, anisotropic temperature factors for proteins have not usually been obtained. The increased resolution possible with synchrotron sources and cryocrystallography will make their determination more common. With this development, it will become possible to obtain better estimates of uncertainties in atom positions than those provided by the Luzzati method.

Publications of refined structures often include a plot of average isotropic B values for side-chain and main-chain atoms of each residue, like that shown in Section III.C, Fig. 8.5b for ALBP. Pictures of the model may be color coded by temperature factor red ("hot") for high values of B and blue ("cold") for low values of B. Either presentation calls the user's attention to parts of the molecule that are vibrationally active and parts that are particularly rigid. Not surprisingly, side-chain temperature factors are larger and more varied (5–60 Å^2) than those of main-chain atoms (5–35 Å^2).

Remember that we see in a crystallographic model an average of all the molecules that diffracted the X rays. Furthermore, we see a static structure representing a stable conformation of a dynamic molecule. It is sobering to realize that the crystallographic model of ALBP exhibits no obvious path for entry and departure of its ligands, which are lipid molecules like oleic acid. Similarly, comparison of the crystallographic models of hemoglobin and deoxyhemoglobin reveals no path for entry of the tiny O_2 molecule. Seemingly simple processes like the binding of small ligands to proteins often involve conformational changes to states not revealed by crystallographic analysis.

Nevertheless, the crystallographic model contributes importantly to solving such problems of molecular dynamics. The refined structure serves as a starting point for simulations of molecular motion. From that starting point, which undoubtedly represents one common conformation of the protein, and from the equations of motion of atoms in the force fields of electrostatic and van der Waals forces, scientists can calculate the normal vibrational motions of the molecules and can simulate random molecular motion, thus gaining insights into how conformational change gives rise to biomolecular function. Even though the crystallographic model is static, it is an essential starting point in revealing the dynamic aspects of structure. As we will see in Chapter 9, time-resolved crystallography offers the potential of giving us highly detailed views of proteins in motion.

D. Other limitations of crystallographic models

The limitations discussed so far apply to all models and suggest questions that the user of crystallographic results should ask routinely. Other limitations are special cases that may or may not apply to a given model. It is important to read the original publications of a structure to see whether any of the following limitations apply.

Low-resolution models

Not all published models are refined to high resolution. For instance, publication of a low-resolution structure may be warranted if it displays an interesting and suggestive arrangement of cofactors or clusters of metal ions, provides possible insights into conformations of a new family or proteins, or displays the application of new imaging methods. In some cases, the published structure is only a crude electron-density model. Or perhaps it contains only the estimated positions of alpha carbons. Such models may be of limited use for comparison with other proteins, but of course, they cannot support detailed molecular analyses. In alpha-carbon models, there is great deal of uncertainty in the positions, and even in the number, of alpha carbons. Often, further refinement of these models reveals errors in the chain tracing. Protein Data Bank header information includes the model resolution and descriptions of its contents. On rare occasions, a low-resolution model file will contain only the coordinates of alpha carbons. Some graphics programs (Chapter 11) will open such files but show no model, giving the impression that the file is empty or damaged. The model appears when the user directs the program to show the alpha-carbon backbone only.

Disordered regions

Occasionally, portions of the known sequence of a protein are never found in the electron-density maps, presumably because the region is highly disordered or in motion, and thus invisible on the time scale of crystallography. It is not at all uncommon for residues at termini, especially the N terminus, to be missing from a model. Discussions of these structure-specific problems are included in a thorough refinement paper, as well as in PDB header information.

Unexplained density

Just as the auto mechanic sometimes has parts left over, electron-density maps occasionally show clear, empty density after all known contents of the crystal have been located. Apparent density can appear as an artifact of missing Fourier terms, but this density disappears when a more complete set of data is obtained. Among the possible explanations for density that is not artifactual are ions like phosphate and sulfate from the mother liquor; reagents like mercaptoethanol, dithiothreitol, or detergents used in purification or crystallization; or cofactors, inhibitors, allosteric effectors, or other small molecules that survived the protein purification. Later discovery of previously unknown but important ligands has sometimes resulted in subsequent interpretation of empty density.

Distortions due to crystal packing

Refinement papers should also mention any evidence that the protein is affected by crystallization. Packing effects may be evident in the model itself. For example, packing may induce slight differences between what are otherwise expected to be identical subunits within an asymmetric unit. Examination of the neighborhood around such differences my reveal that intermolecular contact is a possible cause. In areas where subunits come into direct contact or close contact through intervening water, surface temperature factors are usually lower than at other surface regions.

Functional unit versus asymmetric unit

The symmetry of functional macromolecular complexes in solution is sometimes important to understanding their functions, as in the binding of regulatory proteins having twofold rotational symmetry to palindromic DNA sequences. As discussed in Chapter 4, Section II.H, in the unit cell of a crystal, the largest aggregate of molecules that possesses no symmetry elements, but can be juxtaposed on other identical entities by symmetry operations, is called the *crystallographic asymmetric unit*. Users of models should be careful to distinguish the asymmetric unit from the *functional unit*, which the Protein Data Bank has dubbed the "biologically functional molecule." For example, the functional unit of mammalian hemoglobin is a complex of four subunits, two each of two slightly different polypeptides, called α and β. We say that hemoglobin functions as an $\alpha_2\beta_2$ tetramer. In some hemoglobin crystals, the twofold rotational symmetry axis of the tetramer corresponds to a unit-cell symmetry axis, and the asymmetric unit is a single $\alpha\beta$ dimer. In other cases, the asymmetric unit may contain more than one biological unit.

For technical reasons having to do with data collection strategies, crystal properties, and other processes essential to crystallography itself, the asymmetric unit is often mentioned prominently in papers about new crystallographic models. This discussion is part of a full description of the crystallographic methods for assessment of the work by other crystallographers. It is easy to get the impression that the asymmetric unit is the functional unit, but frequently it is not. Beyond the technical methods sections of a paper, in their interpretations and discussions of the meaning of the model, authors are careful to describe the functional form of the substance under study (if it is known), and this is the form that holds the most interest for users.

It is safe to think of functional-unit symmetry as not necessarily having anything to do with crystallographic symmetry. If the two share some symmetry elements, it is coincidental and may actually be useful to the crystallographer (see, for example, Chapter 7, Section III.C on noncrystallographic symmetry). But in another crystal form of the same substance, the unit cell and the functional unit may share different symmetry elements or none. For you as a user of crystallographic models, looking at crystal symmetry and packing is primarily of value in making sure that you do not make errors in interpreting the model by not allowing for the possibly disruptive effects of crystallization. Once you are confident that the portions of the molecule that pique your interest are not affected by crystal packing, then you can forget about crystal symmetry and the asymmetric unit, and focus on the functional unit.

As mentioned earlier, the asymmetric unit may be only a part of the functional unit. This sometimes poses a problem for users of crystallographic models because the PDB file for such a crystallographic model usually contains only the coordinates of the asymmetric unit. So in the case of hemoglobin, a file may contain only one αβ dimer, which is only half of what the user would like to see. For some molecules of educational or other wide interest, the Protein Data Bank provides files, called Prepared Biologically Functional Molecules, containing the coordinates of all atoms in the functional unit (for example, oxy- and deoxyhemoglobin tetramers are provide as PDB files 1hho and 3hhb). The additional coordinates are computed by aplying symmetry operations to the coordinates of the asymmetric unit. In these models, one can study all the important intersubunit interactions of the full tetramer. Another solution to this problem is for users themselves to compute the coordinates of the additional subunits. Many molecular graphic programs provide for such calculations (Chapter 11, Section III.J). Users should also beware that the asymmetric unit, and hence the PDB file, may contain two or several functional units.

E. Summary

Sensible use of a crystallographic model, like any complex tool, requires an understanding of its limitations. Some limitations, like the precision of atom positions and the static nature of the model, are general constraints on use. Others, like disordered regions, undetected portions of sequence, unexplained density, and packing effects, are model-specific. If you use a protein model from the PDB without reading the header information, or without reading the original publications, you may be missing something vital to the appropriate use of the model. The result may be no more than a crash of your graphics software because of unexpected input like a file containing only alpha carbons. Or more seriously, you may devise and publish a detailed molecular explanation based upon a structural feature that is quite uncertain. In some cases, the model isn't enough. If specific structural details of the model are crucial to a proposed mechanism or explanation, it is advisable to look at the electron-density map in the important region in order to be sure that the map is well defined there and that the model fits it well. The capability to view electron-density maps is already available in some molecular graphics programs for personal computers, and will probably become very common (see Chapter 11).

III. Reading a crystallography paper

A. Introduction

Judging the quality and potential usefulness of a crystallographic model means first extracting the criteria of quality from published reports. To help you begin to develop this skill, I will walk you through an attempt to cull such information from publications of a "typical" crystallographic project. Following are annotated portions of two papers reporting the crystallization and structure determination of adipocyte lipid binding protein (ALBP), a member of a family of hydrophobic-ligand-binding proteins. The first paper[1] appeared in August 1991, announcing the purification and crystallization of the protein, and presenting preliminary results of crystallographic analysis. The second paper[2], which appeared in April 1992, presented the completed structure with experimental details. In the following sections, I will focus primarily on the experimental and results sections of the papers and specifically upon (1) methods and concepts treated earlier in this book and (2) criteria of refinement convergence and quality of the model.

Although I have reproduced parts of the published experimental procedures here (with the permission of the authors and publisher), you may wish to obtain these papers from your library and read them before proceeding with this example. See Footnotes 1 and 2 for complete references.

In the following material, sections taken from the original papers are presented in smaller type. Annotations are in the usual type size. For convenience, figures and tables are renumbered in sequence with those of this chapter. For access to references cited in excerpts, see the complete papers. Stereo illustrations of maps and models (not part of the papers) are derived from files kindly provided by Zhaohui Xu. I am indebted to Xu and to Leonard J. Banaszak for allowing me to use their work as an example and for supplying me with an almost complete reconstruction of this structure determination project.

B. Annotated excerpts of the preliminary (8/91) paper

All reprinted parts of this paper (cited in Footnote 1) appear with the permission of Professor Leonard J. Banaszak and the American Society for Biochemistry and Molecular Biology, Inc., publisher of *Journal of Biological Chemistry*.

[1] Z. Xu, M. K. Buelt, L. J. Banaszak, and G. A. Bernlohr, Expression, purification, and crystallization of the adipocyte lipid binding protein, *J. Biol. Chem.* **266,** 14367–14370, 1991.
[2] Z. Xu, D. A. Bernlohr, and L. J. Banaszak, Crystal structure of recombinant murine adipocyte lipid-binding protein, *Biochemistry* **31,** 3484–3492, 1992.

In the August 5, 1991, issue of *Journal of Biological Chemistry*, Xu, Buelt, Banaszak, and Bernlohr reported the cloning, expression, purification, and crystallization of adipocyte lipid binding protein (ALBP, or rALBP for the recombinant form), along with preliminary results of crystallographic analysis. This type of paper sometimes appears as soon as a research team has carried the structure project far enough to know that it promises to produce a good model. An important aim of announcing that work is in progress on a molecule is to avoid duplication of effort in other laboratories. Although one might cynically judge that such papers constitute a defense of territory, and a grab for priority in the work at hand, something much more important is at stake. Crystallographic structure determination is a massive and expensive undertaking. The worldwide resources, both equipment and qualified scientists, for structure determination are inadequate for the many molecules we would like to understand. Duplication of effort on the same molecule squanders limited resources in this important field. So generally, as soon as a team has good evidence that they can produce a structure, they alert the crystallographic community to prevent parallel work from beginning in other labs. On the other hand, structure determination is becoming much more rapid, and there have been instances of races between labs to solve important structures. For these reasons, preliminary papers are becoming less common.

The following paragraph is an excerpt from the preliminary (8/91) paper, "Experimental Procedures" section:

> *Crystallization*—Small crystals (0.05 × 0.1 × 01 mm) were obtained using the hanging drop/vapor equilibrium method (18). 10-µl drops of 2.5 mg/ml ALBP in 0.05 M Tris, 60% ammonium sulfate, 1 mM EDTA, 1 mM dithiothreitol, 0.05% sodium azide buffer with a pH of 7.0 (crystallization buffer) were suspended over wells containing the same buffer with varying concentrations of ammonium sulfate, from 75 to 85% saturation. Small, well shaped crystals were formed within a month at an 80% saturation and 19°C. These crystals were isolated, washed with mother liquid, and used as seeds by transferring them into a 10-µl drop of 4 mg/ml fresh ALBP in the 80% saturation crystallization buffer over a well containing the same buffer. Large crystals, 0.3 × 0.4 × 0.4 mm, grew in 2 days at a constant temperature of 19°C.

The precipitant used here is ammonium sulfate, which precipitates proteins by salting out. Notice that Xu and co-workers tried a range of precipitant concentrations, probably after preliminary trails over a wider range. Crystals produced by the hanging drop method (Chapter 3, Section III.B) were too small for X-ray analysis but were judged to be of good quality. The small crystals were used as seeds on which to grow larger crystals under the same conditions that produced the best small crystals. This method, called *repeated seeding*, was also discussed in Chapter 3. The initial unseeded crystallization probably fails to produce large crystals because many crystals form at about the same rate, and soluble protein is depleted before any crystals become large. The seeded crystallization is probably effective because it decreases the number of sites of crystal growth, causing more protein to go into fewer crystals. Notice also how much faster crystals grow in the seeded drops (2 days) than in the unseeded (1 month). The preformed crystals provide nucleation sites for immediate further growth, whereas the first crystals form by random nucleation events which are usually rate-limiting in unseeded crystallizations.

> *Data Collection and Processing*—Crystals were analyzed with the area detector diffractometer from Siemens/Nicolet. A 0.8-mm collimator was used, and the crystal to detector distance was set at 12 cm with the detector midpoint at $2\theta = 15°$. One ϕ scan totaling 90° and three Ω scans of 68° with χ at 45° were collected with the Rigaku Ru200 operating at 50 kV and 180 mA. Each frame consisted of a 0.25° rotation taken for 120 s. The diffractometer data were analyzed with the Xengen package of programs (19). Raw data within 50 frames were searched to find about 100 strong reflections which were then indexed, and the cell dimensions were refined by least squares methods. Data from different scans were integrated separately and then merged together.

The angles ϕ, χ, ω, and 2θ refer to the diffractometer angles described in Chapter 4, Section III.D, and shown in Fig. 4.21. The Rigaku Ru200 is the X-ray source, a rotating-anode tube. Each frame of data collection is, in essence, one electronic film on which are recorded all reflections that pass through the sphere of reflection during a 0.25° rotation of the crystal. This rotation size is chosen to collect as many reflections as possible without overlap. As mentioned in Chapter 4, diffractometer measurements are almost fully automated. In this instance, cell dimensions were worked out by a computer program that finds 100 strong reflections and indexes them. Then the program employs a least-squares routine (Chapter 7, Section VI.A) to refine the unit-cell dimensions, by finding the cell lengths and angles that minimize the difference between the actual positions of the 100 test reflections and the positions of the same reflections as calculated from the current trial set of cell dimensions. (Least-squares procedures are used in many areas of crystallography in addition to structure refinement.) Using accurate cell dimensions, the program indexed all reflections, and then integrated the X-ray counts received at each location to obtain reflection intensities.

The following excerpt is from the "Results and Discussion" section of the 9/91 paper:

> Crystallization experiments using rALBP were immediately successful. With seeding, octahedral crystals of the *apo*-protein grew to a length of 0.4 mm and a height of 0.3 mm. These crystals give diffraction data to 2.4 Å. An entire data set was collected to 2.7-Å resolution using the area detector system. Statistical details of the combined X-ray data set are presented in Table 8.1.

Table 8.1

X-Ray Data Collection Statistics for Crystalline ALBP

Merging R-factor based on I	0.0426
Resolution limits	2.2 Å
Number of observations	20,478
Number of unique x-ray reflections collected	5,473
Average number of observations for each reflection	4.0 Å
% of possible reflections collected to 2.7 Å	98
% of possible reflections collected to 2.4 Å	36

From Z. Xu *et al.* (1991) *J. Biol. Chem.* **266**, pp. 14367–14370, with permission.

Xu and colleagues had exceptionally good fortune in obtaining crystals. Efforts to crystallize a desirable protein can give success in a few weeks, or never, or anything in between. The time required here exceeds even very optimistic hopes. The extent of diffraction in preliminary tests (2.4 Å) is a key indicator that the crystals might yield a high-quality structure.

Table 8.1 provides you with a glimpse into the quality of the native data set. The 0.25° frames of data from the area detector are merged into one data set by multiplying all intensities in each frame by a scale factor. A least-squares procedure determines scale factors that minimize the differences between intensities of identical reflections observed on different frames. The merging R-factor [see Eq. (7.10)] gives the level of agreement among the different frames of data after scaling. In this type of R-factor, $|\mathbf{F}_{obs}|$s are derived from averaged, scaled intensities for all observations of one reflection, and corresponding $|\mathbf{F}_{calc}|$s are derived from scaled intensities for individual observations of the same reflection. The better the agreement between these two quantities throughout the data set, the lower the merging R-factor. In this case, individual scaled intensities agree with their scaled averages to within about 4%.

You can see from Table 8.1 that 98% of the reflections available out to 2.7 Å [those lying within a sphere of radius $1/(2.7$ Å$)$ centered at the origin of the reciprocal lattice] were measured, and on the average, each reflection was measured four times. Additional reflections were measured out to 2.4 Å. The number of available reflections increases with the third power of the radius of the sampled region in the reciprocal lattice (because the volume of a sphere of radius r is proportional to r^3), so a seemingly small increase in resolution from 2.7 to 2.4 Å requires 40% more data. [Compare $(1/2.4)^3$ with $(1/2.7)^3$.] For a rough calculation of the number of available reflections at specified resolution, see annotations of the 4/92 paper.

The lattice type was orthorhombic with unit cell dimension of $a = 34.4$ Å, $b = 54.8$ Å, $c = 76.3$ Å. The X-ray diffraction data were examined for systematic absences to determine the space group. Such absences were observed along the $\mathbf{a}^*$, $\mathbf{b}^*$, and $\mathbf{c}^*$ axes. Only reflections with h, k, or $l = 2n$ were observed along the reciprocal axes. This indicated that the space group is $P2_12_12_1$ (25). A unit cell with the dimensions described above has a volume of 1.44×10^5 Å^3. Assuming that half of the crystal volume is water, the volume of protein is approximately 7.2×10^4 Å^3. Considering the space group here, the volume of protein in 1 asymmetric unit would be 1.8×10^4 Å^3. By averaging the specific volume of constituent amino acids, the specific volume of ALBP is 0.715 mL/g. This led to the conclusion that the molecular mass in one asymmetric unit is 15,155 daltons. Since the molecular mass of ALBP is approximately 15 kDa, there is only 1 molecule of ALBP in an asymmetric unit.

Recall from Chapter 5, Section IV.C, that for a twofold screw axis along the $\mathbf{c}$ edge, all odd-numbered $00l$ reflections are absent. In the space group $P2_12_12_1$, the unit cell possesses twofold screw axes on all three edges, so odd-numbered reflections on all three principle axes of the reciprocal lattice ($h00$, $0k0$, and $00l$) are missing. The presence of only even-numbered reflections on the reciprocal-lattice axes announces that the ALBP unit cell has $P2_12_12_1$ symmetry.

As described in Chapter 3, Section IV, the number of molecules per asymmetric unit can be determined from unit-cell dimensions and a rough estimate of the protein/water ratio. Since this number is an integer, even a rough calculation can give a reliable answer. The assumption that ALBP crystals are 50% water is no more than a guess taken from near the middle of the range for protein crystals (30 to 78%). The unit-cell volume is $(34.4$Å$)(54.8$Å$)(76.3$Å$) = 1.44 \times 10^5$ Å^3, and if half that volume is protein, the protein volume is 7.2×10^4 Å^3. In space group $P2_12_12_1$, there are four equivalent positions (Chapter 4, Section II.H), so there are four asymmetric units per unit cell. Each one must occupy one-fourth of the protein volume, so the volume of the asymmetric unit is one-fourth of 7.2×10^4, or 1.8×10^4 Å^3. The stated specific volume (volume per gram) of the protein is the weighted average of the specific volumes of the amino acid residues (which can be looked up), weighted according to the amino-acid composition of ALBP. The molecular mass of one asymmetric unit is obtained by converting the density of ALBP in grams per milliliter (which is roughly the inverse of the specific volume) to daltons per cubic angstrom, and then multiplying by the volume of the asymmetric unit, as follows:

$$\frac{1\text{g}}{0.715\text{mL}} \cdot \frac{1 \text{ ml}}{\text{cm}^3} \cdot \frac{\text{cm}^3}{(10^8)^3 \text{ Å}_3} \cdot \frac{6.02 \times 10^{23} \text{ daltons}}{\text{g}} \cdot 1.8 \times 10^4 \text{ Å}^3$$

$$= 1.5 \times 10^4 \text{ daltons.}$$

This result is very close to the known molecular mass of ALBP, so there is one ALBP molecule per asymmetric unit. This knowledge is an aid to early map interpretation.

> As indicated, ALBP belongs to a family of low molecular weight fatty acid binding proteins. The sequences of the proteins in the family have been shown to be very similar and in particular in the amino-terminal domain where Y19[3] resides. Among them, the structure of myelin P2 and IFABP has been solved. Since the amino acid identity between ALBP and myelin P2 is about 69%, P2 should be a good starting structure to obtain phase information for ALBP using the method of molecular replacement. Preliminary solutions to the rotation and translation functions have been obtained. Seeding techniques will allow us to obtain large crystals for further study of the *holo-* and phosphorylated protein. By comparing the crystal structures of these different forms, it should be possible to structurally determine the effects of protein phosphorylation on ligand binding and ligand binding on phosphorylation.

Because ALBP is related to several proteins of known structure, molecular replacement is an attractive option for phasing. The choice of a phasing model is simple here: just pick the one with the amino-acid sequence most similar to ALBP, which is myelin P2 protein. Solution of rotation and translation functions refers to the search for orientation and position of the phasing model (P2) in the unit cell of ALBP. The subsequent paper provides more details.

C. Annotated excerpts from the full structure-determination (4/92) paper

All reprinted parts of this paper (cited in Footnote 2) appear with the permission of Professor Leonard J. Banaszak and the American Chemical Society, publisher of *Biochemistry*.

In April 1992, the structure determination paper appeared in *Biochemistry*. This paper contains a full description of the experimental work, and a complete analysis of the structure. The following is from the 4/92 paper, "Abstract" section:

> Adipocyte lipid-binding protein (ALBP) is the adipocyte member of an intracellular hydrophobic ligand-binding protein family. ALBP is phosphorylated by the insulin receptor kinase upon insulin stimulation. The crystal structure of recombinant murine ALBP has been determined and refined to 2.5 Å. The final *R*-factor for the model is 0.18 with good canonical properties.

A 2.5-Å model refined to an *R*-factor of 0.18 should be a detailed model. "Good canonical properties" means good agreement with accepted values of bond lengths, bond angles, and planarity of peptide bonds.

---

[3]Y19 is tyrosine 19, a residue considered important to the function of ALBP.

The following excerpts are from the "Materials and Methods" section of the 4/92 paper:

> *Crystals and X-ray Data Collection.* Detailed information concerning protein purifica-
> tion, crystallization, and X-ray data collection can be found in a previous report (Xu
> *et al.*, 1991) and will be mentioned here in summary form. Recombinant murine *apo*-
> ALBP crystallizes in the orthorhombic space group $P2_12_12_1$ with the following unit
> cell dimensions $a = 34.4$ Å, $b = 54.8$ Å, and $c = 76.3$ Å. The asymmetric unit con-
> tains one molecule with a molecular weight of 14,500. The entire diffraction data set
> was collected on one crystal. In the resolution range ∞–2.5 Å, 5115 of the 5227 theo-
> retically possible reflections were measured. Unless otherwise noted the diffraction
> data with intensities greater than 2σ were used for structure determination and refine-
> ment. As can be seen in Table 8.2, this included about 96% of the measured data.

This section reviews briefly the results of the preliminary paper. In the early stages of the work, reflections weaker than two times the standard deviation for all reflections (2σ) were omitted from Fourier syntheses, because of greater un-certainty in the measurements of weak reflections. Table 8.2 is discussed later.

The diffractometer software computes the number of reflections available at 2.5-Å resolution by counting the number of reciprocal-lattice points that lie within a sphere of radius $[1/(2.5\text{Å})]$, centered at the origin of the reciprocal lattice. This number is roughly equal to the number of reciprocal unit cells within the $1/(2.5\text{Å})$ sphere, which is, again roughly, the volume of the sphere (V_{rs}) divided by the volume of the reciprocal unit cell (V_{rc}). The volume of the reciprocal unit cell is the inverse of the real unit-cell volume V. So the number of reflections available at 2.5-Å resolution is approximately $(V_{rs}) \cdot (V)$. Because of the symmetry of the reciprocal lattice and of the $P2_12_12_1$ space group, only one-eighth of the reflections are unique (Chapter 4, Section III.G). So the number of unique reflections is approximately $(V_{rs}) \cdot (V)/8$, or

$$\frac{\frac{4}{3}\pi\left(\frac{1}{2.5\text{Å}}\right)^3 (1.44 \times 10^5 \text{Å}^3)}{8} = 4825 \text{ reflections.}$$

The 8% difference between this result and the stated 5227 reflections is due to the approximations made here and to the sensitivity of the calculation to small round-off in unit-cell dimensions.

> *Molecular Replacement.* The tertiary structure of crystalline ALBP was solved by
> using the molecular replacement method incorporated into the XPLOR computer pro-
> gram (Brunger *et al.,* 1987). The refined crystal structure of myelin P2 protein without
> solvent and fatty acid was used as the probe structure throughout the molecular re-
> placement studies. We are indebted to Dr. A. Jones and his colleagues for permission
> to use their refined P2 coordinates before publication.

Note that the myelin P2 coordinates were not yet available from the Protein Data Bank and were obtained directly from the laboratory in which the P2 structure was determined. Because of the time required for publication of research papers and processing of coordinates by the PDB, coordinates may be available directly from a crystallographic research group several months before they are available from PDB.

In this project, the search for the best orientation and position of P2 in the ALBP unit cell was divided into three parts: a rotation search to find promising orientations, refinement of the most promising orientations to find the best orientation, and a translation search to find the best position. Here are the details of the search:

> (1) *Rotation Search.* The rotation search was carried out using the Patterson search procedures in XPLOR. The probe Patterson maps were computed from structure factors calculated by placing the P2 coordinates into an orthorhombic cell with 100-Å edges. One thousand highest Patterson vectors in the range of 5–15 Å were selected and rotated using the pseudoorthogonal Eulerian angles (θ_+, θ_2, θ_-) as defined by Lattman (1985). The angular search interval for θ_2 was set to 2.5°; intervals for θ_+ and θ_- are functions of θ_2. The rotation search was restricted to the asymmetric unit $\theta_- = 0\text{--}180°$, $\theta_2 = 0\text{--}90°$, $\theta_+ = 0\text{--}720°$ for the $P2_12_12_1$ space group (Rao, *et al.*, 1980). XPLOR produces a sorted list of the correlation results simplifying final interpretation (Brunger 1990).

XPLOR is a modern package of refinement programs that includes powerful procedures for energy refinement by simulated annealing, in addition to more traditional tools like least-squares methods and molecular-replacement searches. The package is available for use on many different computer systems. Simulated annealing for large molecules usually requires supercomputers.

The P2 phasing model is referred to here as the *probe*. For the rotation search, the probe was placed in a unit cell of arbitrary size and $\mathbf{F}_{calc}$s were obtained from this molecular model, using Eq. (5.15). Then a Patterson map was computed from these $\mathbf{F}_{calc}$s using Eq. (6.10). Recall that Patterson maps reflect the molecule's orientation but not its position. All peaks in the Patterson map except the strongest 1000 were eliminated. Then the resulting simplified map was compared to a Patterson map calculated from ALBP reflection intensities. The probe Patterson was rotated in a three-dimensional coordinate system to find the orientation that best fit the ALBP Patterson. (The angles refer to a standard set of angles for rotating the model through all unique orientations.) A plot of the angles versus some criterion of coincidence between peaks in the two Patterson maps is called a *rotation function*. Peaks in the rotation function occur at sets of angles where many coincidences occur. The coincidences are not perfect because there is a finite interval between angles tested, and the exact desired orientation is likely to lie between test angles. The interval is made small enough to avoid missing promising orientations altogether.

(2) *Patterson Correlation Refinement.* To select which of the orientations determined from the rotation search is the correct solution a Patterson correlation refinement of the peak list of the rotation function was performed. This was carried out by minimization against a target function defined by Brunger (1990) and as implemented in XPLOR. The search model, P2, was optimized for each of the selected peaks of the rotation function.

As discussed later in the "Results" section, the rotation function contains many peaks. The strongest 100 peaks are selected, and each orientation is refined by least squares to produce the best fit to the ALBP Patterson map. For each refined orientation, a correlation coefficient is computed. The orientation giving the highest correlation coefficient is chosen as the best orientation for the phasing model.

(3) *Translation Search.* A translation search was done by using the P2 probe molecule oriented by the rotation function studies and refined by the Patterson correlation method. The translation search employed the standard linear correlation coefficient between the normalized observed structure factors and the normalized calculated structure factors (Funinaga & Read, 1987; Brunger, 1990). X-ray diffraction data from $10-3-\text{Å}$ resolution were used. Search was made in the range $x = 0–0.5$, $y = 0–0.5$, and $z = 0–0.5$, with the sampling interval 0.0125 of the unit cell length.

The last step in molecular replacement is to find the best position for the probe molecule in the ALBP unit cell. The P2 orientation obtained from the rotation search and refinement is tried in all unique locations at intervals of one-eightieth (0.0125) of the unit-cell axis lengths. The symmetry of the $P2_12_12_1$ unit cell allows this search to be confined to the region bound by one-half of each cell axis. The total number of positions tested is thus (40)(40)(40) or 64,000. For each position, F_{calc}s are computed [Eq. (5.15)] from the P2 model and their amplitudes are compared with the $|F_{obs}|$s from the ALBP native data set. An unspecified correlation coefficient, probably similar to an R-factor, is computed for each P2 position, and the position giving P2 $|F_{calc}|$s in best agreement with ALBP $|F_{obs}|$s is chosen as the best position for P2 as a phasing model. The starting phase estimates for the refinement were thus the phases of F_{calc}s computed [Eq. (5.15)] from P2 in the final orientation and position determined by the three-stage molecular-replacement search.

Structure Refinement. The refinement of the structure was based on an energy function approach (Brunger *et al.*, 1987): arbitrary combinations of empirical and effective energy terms describing crystallographic data as implemented in XPLOR. Molecular model building was done on an IRIS Workstation (Silicon Graphics) with the software TOM, a version of FRODO (Jones, 1978).

The initial model of ALBP was built by simply putting the amino acid sequence of ALBP into the molecular structure of myelin P2 protein. After a 20-step rigid-body refinement of the positions and orientations of the molecule, crystallographic refinement with simulated annealing was carried out using a slow-cooling protocol (Brunger *et al.*, 1989, 1990). Temperature factor refinement of grouped atoms, one for backbone and one for side-chain atoms for each residue, was initiated after the R -factor dropped to 0.249.

The first electron-density map was computed [Eq. (7.03)] with $|F_{obs}|$s from the ALBP data set and α_{calc}s from the oriented P2 molecule. Plate 13 shows a small section of this map superimposed upon the *final* model.

An early map like Plate 13, computed from initial phase estimates, harbors many errors, where the map does not agree with the model ultimately derived from refinement. In this section, you can see both false breaks and false connections in the density. For example, there are breaks in density at C_β of the phenylalanine residue (side chain ending with six-membered ring) on the right, and along the protein backbone at the upper left. The lobe of density corresponding to the valine side chain (center front) is disconnected and out of place. There is a false connection between density of the carbonyl oxygen (red) at lower left and side chain density above. Subsequent refinement is aimed at improving this map.

Next, the side chains of P2 were replaced with the side chains of ALBP at corresponding positions in the amino-acid sequence to produce the first ALBP model. The position and orientation of this model were refined by least squares, treating the model as a rigid body. Subsequent refinement was by simulated annealing. At first, all temperature factors were constrained at 15.0 $Å^2$. After the first round of simulated annealing, temperature factors were allowed to refine for atoms in groups, one value of B for all backbone atoms within a residue and another for side-chain atoms in the residue.

> The new coordinates were checked and adjusted against a $(2|F_o| - |F_c|)$ and a $(|F_o| - |F_c|)$ electron density map, where $|F_o|$ and $|F_c|$ are the observed and calculated structure factor amplitudes. Phases are calculated from the crystal coordinates. The Fourier maps were calculated on a grid corresponding to one-third of the high-resolution limit of the input diffraction data. All residues were inspected on the graphics system at several stages of refinement. The adjustments were made on the basis of the following criteria: (a) that an atom was located in low electron density in the $(2|F_o| - |F_c|)$ map or negative electron density in the $(|F_o| - |F_c|)$ map; (b) that the parameters for the Φ, Ψ angles placed the residue outside the acceptable regions in the Ramachandran diagram. Iterative refinement and model adjustment against a new electron density map was carried out until the R-factor appeared unaffected. Isotropic temperature factors for individual atoms were then included in the refinement.

In between rounds of computerized refinement, maps were computed using $|F_{obs}|$s from the ALBP data set and α_{calc}s from the current model [taken from $|F_{calc}|$s computed by Eq. (5.15)]. The model was corrected where the fit to maps was poor, or where the Ramachandran angles Φ and Ψ were forbidden. Notice that the use of $2F_o - F_c$ and $F_o - F_c$ maps [Eqs. (7.4) and (7.5)] is as described in Chapter 7, Section IV.B. When alternating rounds of refinement and map fitting produced no further improvement in R-factor, temperature factors for each atom were allowed to refine individually, leading to further decrease in R.

> The next stage of the crystallographic study included the location of solvent molecules. They were identified as well-defined peaks in the electron-density maps within hydrogen-bonding distance of appropriate protein atoms or other solvent atoms. Solvent atoms were assigned as water molecules and refined as oxygen atoms. Those that refined to positions too close to other atoms, ended up located in low electron density, or had associated temperature factors greater than 50 Å^2 were removed from the coordinate list in the subsequent stage. The occupancy for all atoms, including solvent molecules, was kept at 1.0 throughout the refinement. Detailed progress of the crystallographic refinement is given in Table 8.2.

Finally, ordered water molecules were added to the model where unexplained electron-density was present in chemically feasible locations for water molecules. Temperature factors for these molecules (treated as oxygen atoms) were allowed to refine individually. If refinement moved these molecules into unrealistic positions or increased their temperature factors excessively, the molecules were deleted from the model. Occupancies were constrained to 1.0 throughout the refinement. This means that B values reflect both thermal motion and disorder (Section II.C). Because all B values fall into a reasonable range, the variation in B can be attributed to thermal motion. Table 8.2 shows the progress of the refinement.

Note that R drops precipitously in the first stages of refinement after ALBP side chains replace those of P2. Note also that R and the deviations from ideal bond lengths, bond angles, and planarity of peptide bonds decline smoothly throughout the later stages of refinement. The small increase in R at the end is due to inclusion of weaker reflections in the final round of simulated annealing.

The following excerpts are from the "Results" section of the 8/92 paper:

> *Molecular Replacement.* From the initial rotation search, the 101 highest peaks were chosen for further study. These are shown in Fig. 8.4. The highest peak of the rotation function had a value 4.8 times the standard deviation above the mean and 1.8 times the standard deviation above the next highest peak. The orientation was consistently the highest peak for diffraction data within the resolution ranges 10–5, 7–5, and 7–3 Å. Apart from peak number 1, six strong peaks emerged after PC[4] refinement, as can be seen in Fig. 8.4b. These peaks all corresponded to approximately the same orientation as peak number 1. Three of them were initially away from that orientation and converged to it during the PC refinement.
>
> A translation search as implemented in XPLOR was used to find the molecular position of the now oriented P2 probe in the ALBP unit cell. Only a single position

---

[4]Patterson correlation.

Table 8.2

Progress of Refinement

Stage[a]	Number of reflections	R-factor	B (Å^2)	Solvent included	RMS deviations Bond length (Å)	Bond angle (deg)	Planarity (deg)
1	2976	0.458	15.0		0.065	4.12	9.015
2	2976	0.456	15.0		0.065	4.12	9.012
3	4579	0.235	Group		0.019	3.17	1.506
4							
5	4579	0.220	Indiv.		0.018	3.77	1.408
6							
7	4579	0.197	Indiv.	31	0.018	3.73	1.366
8							
9	4579	0.172	Indiv.	88	0.016	3.47	1.139
10							
11	4773	0.183	Indiv.	69	0.017	3.46	1.070

Reprinted with permission from Z. Xu *et al.* (1992). *Biochemistry* **31**, 3484–3492. Copyright 1992 American Chemical Society.

[a] Key to stages of refinement:

Stage	Action
1	Starting model
2	Rigid-body refinement
3	Simulated annealing
4	Model rebuilt using ($2\mathbf{F_o}-\mathbf{F_c}$) and $\mathbf{F_o}-\mathbf{F_c}$) electron-density maps
5	Simulated annealing
6	Model rebuilt using ($2\mathbf{F_o}-\mathbf{F_c}$) and $\mathbf{F_o}-\mathbf{F_c}$) electron-density maps, H_2O included
7	Simulated annealing
8	Model rebuilt using ($2\mathbf{F_o}-\mathbf{F_c}$) and $\mathbf{F_o}-\mathbf{F_c}$) electron-density maps, H_2O included
9	Simulated annealing
10	Model rebuilt using ($2\mathbf{F_o}-\mathbf{F_c}$) and $\mathbf{F_o}-\mathbf{F_c}$) electron-density maps, H_2O included
11	Simulated annealing

emerged at $x = 0.250$, $y = 0.425$, $z = 0.138$ with a correlation coefficient of 0.419. The initial R-factor for the P2 coordinates in the determined molecular orientation and position was 0.470 including X-ray data in the resolution range of 10–3 Å. A rigid-body refinement of orientation and position reduced the starting R-factor to only 0.456, probably attesting to the efficacy of the Patterson refinement in XPLOR.

In Figure 8.4a, the value of the rotation function, which indicates how well the probe and ALBP Patterson maps agree with each other, is plotted vertically against numbers assigned to the 101 orientations that produced best agreement. Then each of the 101 orientations were individually refined further, by finding the nearby orientation having maximum value of the rotation function. In some cases, different peaks refined to the same final orientation.

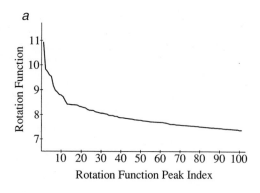

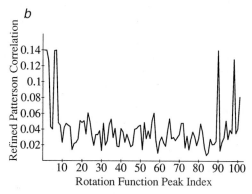

Figure 8.4   Rotation function results: P2 into crystalline ALBP. (*a*) Plot of the 101 best solutions to the rotation function, each peak numbered in the horizontal direction (abscissa). The correlation between the Patterson's of the probe molecule and the measured ALBP X-ray results are shown in the vertical direction (ordinate) and are given in arbitrary units. (*b*) Description of the rotation studies after Patterson correlation refinement. The peak numbers plotted in both panels (*a*) and (*b*) are the same. Reprinted with permission from Z. Xu *et al.* (1992) *Biochemistry* **31**, 3484–3492. Copyright 1992 American Chemical Society.

Each refined orientation of the probe received a correlation coefficient that shows how well it fits the Patterson map of ALBP. The orientation receiving the highest correlation coefficient was taken as the best orientation of the probe, and then used to refine the position of the probe in the ALBP unit cell. The orientation and position of the model obtained from the molecular replacement search was so good that refinement of the model as a rigid body produced only slight improvement in *R*. The authors attribute this to the effectiveness of the Patterson correlation refinement of model orientation, stage two of the search.

Refined Structure of apo-ALBP. The refined ALBP structure has a *R*-factor 0.183 when all observed X-ray data (4773 reflections) between 8.0 and 2.5 Å are included. The rms deviation of bond lengths, bond angle, and planarity from ideality is 0.017 Å, 3.46°, and 1.07°, respectively. An estimate of the upper limit of error in atomic coordinates is obtained by the method of Luzzati (1952). Fig. 8.5 summarizes the overall refined model.

The plot presented in Fig. 8.5*a* suggests that the upper limit for the mean error of the refined ALBP coordinates is around 0.34 Å. The mean temperature factors for main-chain and side-chain atoms are plotted in Fig. 8.5*b*.

The final *R*-factor and structural parameters exceed the standards described in Section I and attest to the high quality of this model. Atom locations are precise to an average of 0.34 Å, about one-fifth of a carbon–carbon covalent bond length. The plot of temperature factors shows greater variability and range for side-chain atoms, as expected, and shows no outlying values. The model defines the positions of all amino-acid residues in the protein.

Careful examination of $(2|\mathbf{F}_o|-|\mathbf{F}_c|)$ and $(|\mathbf{F}_o|-|\mathbf{F}_c|)$ maps at each refinement step led to the conclusion that no bound ligand was present. There was no continuous positive electron density present near the ligand-binding site as identified in both P2 (Jones, *et al.*, 1988) and IFABP (Sacchettini *et al.*, 1989a). The absence of bound fatty acid in crystalline ALBP is consistent with the chemical modification experiment which indicates ALBP purified from *E. coli* is devoid of fatty acid (Xu *et al.*, 1991). The final refined coordinate list includes 1017 protein atoms and 69 water molecules.

The final maps exhibit no unexplained electron density. This is of special concern because ALBP is a ligand-binding protein (its ligand is a fatty acid), and ligands sometimes survive purification and crystallization, and are found in the final electron-density map. It is implied by references to *apo*-protein and *holo*-protein that attempts to determine the structure of an ALBP-ligand complex are under way. If it is desired to detect conformational changes upon ligand binding, then it is crucial to know that no ligand is bound to this *apo*-protein, so that conformational differences between *apo*- and *holo*-forms, if found, can reliably be attributed to ligand binding.

To compare *apo*- and *holo*-forms of proteins after both structures have been determined independently, crystallographers often compute difference Fourier syntheses (Chapter 7, Section IV.B), in which each Fourier term contains the structure-factor difference $\mathbf{F}_{holo}-\mathbf{F}_{apo}$. A contour map of this Fourier series is called a difference map, and it shows only the differences between the *holo*- and *apo*- forms. Like the $\mathbf{F}_o-\mathbf{F}_c$ map, the $\mathbf{F}_{holo}-\mathbf{F}_{apo}$ map contains both positive and negative density. Positive density occurs where the electron density of the *holo*-form is greater than that of the *apo*-form, so the ligand shows up clearly in positive density. In addition, conformational differences between *holo*- and *apo*-forms result in positive density where *holo*-protein atoms occupy regions that are unoccupied in the *apo*-form, and negative

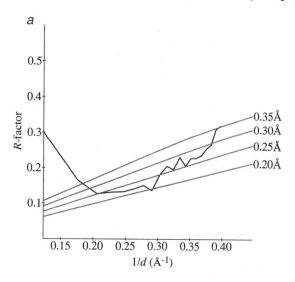

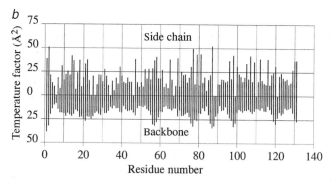

Figure 8.5 ALBP refinement results. (*a*) Theoretical estimates of the rms positional errors in atomic coordinates according to Luzzati (1952) are shown superimposed on the curve for the ALBP diffraction data. The coordinate error estimated from this plot is 0.25 Å with an upper limit of about 0.35 Å. (*b*) Mean values of the main-chain and side-chain temperature factors are plotted versus the residue number. The temperature factors are those obtained from the final refinement cycles. Reprinted with permission from Z. Xu *et al.* (1992) *Biochemistry* **31**, 3484–3492. Copyright 1992 American Chemical Society.

density where *apo*-protein atoms occupy regions that are unoccupied in the *holo*-form. The standard interpretation of such a map is that negative density indicates positions of protein atoms before ligand binding, and positive density locates the same atoms after ligand binding. In regions where the two forms are identical, $\mathbf{F}_{holo} - \mathbf{F}_{apo} = 0$, and the map is blank.

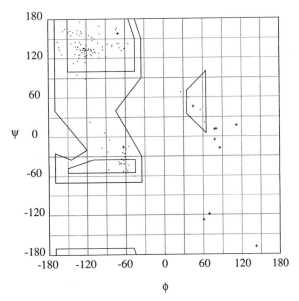

Figure 8.6 Ramachandran plot of the crystallographic model of ALBP. The main-chain torsional angle Φ (N–C$_\alpha$ bond) is plotted versus Ψ (C–C$_\alpha$ bond). The following symbols are used: (Å) nonglycine residues; (+) glycine residues. The enclosed areas of the plot show sterically allowed angles. Reprinted with permission from Z. Xu *et al.* (1992) *Biochemistry* **31**, 3484–3492. Copyright 1992 American Chemical Society.

Structural Properties of Crystalline ALBP. A Ramachandran plot of the main chain dihedral angle Φ and Ψ is shown in Fig. 8.6. In the refined model, 13 residues have positive
Φ angles, 9 of which belong to glycine residues. There are 11 glycine residues in ALBP,
all associated with good quality electron density.

Most of the residues having forbidden values of Φ and Ψ are glycines, represented by "+" in Fig. 8.6, whereas all other amino acids are represented by dots. Succeeding discussion reveals that these unusual conformations are also found in P2 and other members of this protein family, strengthening the argument that these conformations are not errors in the model, and suggesting that they might be important to structure and/or function in this family of proteins.

Plate 2 shows, at the end of refinement, the same section of map as in Plate 13. By comparing Plates 2 and 13, you can see that the map errors described earlier were eliminated, and that the map is a snug fit to a chemically, stereochemically, and conformationally realistic model.

IV. Summary

All crystallographic models are *not* equal. The noncrystallographer can assess model quality by carefully reading original publications of a macromolecular structure. The kind of reading and interpretation implied by my annotations in Section II is essential to wise use of models. Don't get me wrong; there is no attempt on the part of crystallographers to hide the limitations of models. On the contrary, refinement papers often represent almost heroic efforts to make plain what the final model says and leaves unsaid. These efforts are in vain if the reader does not understand them, or worse, never reads them. These efforts are often undercut by the simple power of the visual model. The brightly colored stereo views of a protein model, which are in fact more akin to cartoons than to molecules, endow the model with a concreteness that exceeds the intentions of the thoughtful crystallographer. It is impossible for the crystallographer, with vivid recall of the massive labor that produced the model, to forget its shortcomings. It is all too easy for users of the model to be unaware of them. It is also all too easy for the user to be unaware that, through temperature factors, occupancies, undetected parts of the protein, and unexplained density, crystallography reveals more than a single molecular model shows.

   Even the highest-quality model does not explain itself. If I showed you a perfect model of a protein of unknown function, it is highly unlikely that you could tell me what it does, or even pinpoint the chemical groups critical to its action. Using a model to explain the properties and action of a protein means bringing the model to bear upon all the other available evidence. This involves gaining intimate knowledge of the model, a task roughly as complex as learning your way around a small city. In Chapter 11, I will discuss the exploration of macromolecular models by computer graphics. But first, I must carry out two other tasks. In the next chapter, I will build on your understanding of X-ray diffraction to introduce you to other means of structure determination using diffraction, including X-ray diffraction of fibers and powders, and diffraction by neutrons and electrons. Then, in Chapter 10, I will briefly introduce other, noncrystallographic, methods of structure determination, in particular NMR and homology modeling. With each method, I will discuss how to assess the quality of the resulting models, by analogy with the criteria covered in this chapter.

9

Other Diffraction Methods

I. Introduction

The same principles that underlie single-crystal X-ray crystallography make other kinds of diffraction experiments understandable. In this chapter, I provide brief, qualitative descriptions of other diffraction methods. First is X-ray diffraction by fibers rather than crystals. Next is diffraction by amorphous materials like powders and solutions. Then I will look at diffraction using other forms of radiation, specifically neutrons and electrons. I will show how each type of diffraction experiment provides structural information that can complement or supplement information from single-crystal X-ray diffraction. Finally, returning to X rays and crystals, I will discuss Laue diffraction, which allows the collection of a full diffraction data set from a single brief exposure of a crystal to polychromatic x-radiation. Laue diffraction opens the door to time-resolved crystallography, yielding crystallographic models of intermediate states in chemical reactions. For each of the crystallographic methods, I provide references to recent research articles that exemplify the method.

II. Fiber diffraction

Many important biological substances do not form crystals. Among these are most membrane proteins and fibrous materials like collagen, DNA, filamentous viruses, and muscle fibers. Some membrane proteins can be crystallized in matrices of lipid and studied by X-ray diffraction (Chapter 3, Section III.D), or they can be incorporated into lipid films (which are in essence two-dimensional crystals) and studied by electron diffraction. I will discuss electron diffraction later in this chapter. Here I will examine diffraction by fibers.

Like crystals, fibers are composed of molecules in an ordered form. When irradiated by an X-ray beam perpendicular to the fiber axis, fibers produce distinctive diffraction patterns that reveal their dimensions at the molecular level. Because many fibrous materials are polymeric and of known chemical composition and sequence, their molecular dimensions are sometimes all that is needed to build a feasible model of their structure.

Some materials (for example, certain muscle proteins) form fibers spontaneously or are naturally found in fibrous form. Many other polymeric substances, like DNA, can be induced into fibers by pulling them from an amorphous gel with tweezers or a glass rod. For data collection, the fiber is simply suspended between a well-collimated X-ray source and a detector, such as film (Fig. 9.1).

The order in a fiber is one-dimensional (along the fiber) rather than three-dimensional, as in a crystal. You can think of molecules in a fiber as being stretched out parallel to the fiber axis but having their termini occurring at random along the fiber, as shown in the expanded detail of the fiber in Fig. 9.1. Because the X-ray beam simultaneously sees all molecules in all possible rotational orientations about the fiber axis, Bragg reflections from a fiber are cylindrically averaged, and irradiation of the fiber by a beam perpendicular to the fiber axis gives a complete, but complex, diffraction pattern from a single orientation of the fiber.

Fibers can be crystalline or noncrystalline. Crystalline fibers are actually composed of long, thin microcrystals oriented with their long axis parallel to the fiber axis. When a crystalline fiber is irradiated with X rays perpendicular to the fiber axis, the result is the same as if a single crystal were rotated about its axis in the X-ray beam during data collection, sort of like a rotation photograph taken over 360° instead of the usual very small rotation angle (Chapter 4, Section III.D). All Bragg reflections are registered at once, on *layer lines* perpendicular to the fiber axis. All fiber diffraction patterns have two mirror planes, parallel (the meridian) and perpendicular (the equator) to the fiber axis. Many of the reflections overlap, making analysis of the diffraction pattern very difficult.

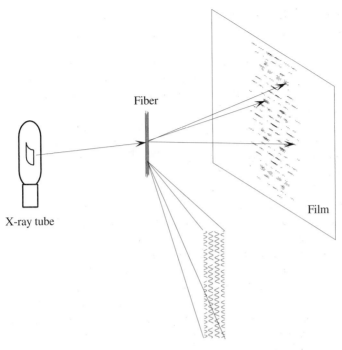

Figure 9.1 Fiber diffraction. Molecules in the fiber are oriented parallel to the beam axis but aligned at random. X-ray beams emerging from the fiber strike the detector in layer lines perpendicular to the fiber axis.

In noncrystalline fibers, all the *molecules* (as opposed to oriented groups of them in microcrystals) are parallel to the fiber axis but aligned along the axis at random. This arrangement gives a somewhat simpler diffraction pattern, also consisting of layer lines, but with smoothly varying intensity rather than distinct reflections. In the diffraction patterns from both crystalline and noncrystalline materials, spacings of layer lines are related to the periodicity of the individual molecules in the fiber, as I will show.

A simple and frequently occurring structural element in fibrous materials is the helix. I will use the relationship between the dimensions of simple helices and that of their diffraction patterns to illustrate how diffraction can reveal structural information. As a further simplification, I will assume that the helix axis is parallel to the fiber axis. As in all diffraction methods, the diffraction pattern is a Fourier transform of the object in the X-ray beam, averaged over all the orientations present in the sample. In the case of fibers, this means that the transform is averaged cylindrically, around the molecular axis parallel to the fiber axis.

Figure 9.2 shows some simple helices and their transforms. The transform of the helix in Fig. 9.2a exhibits an X pattern that is always present in transforms of helices. I will explain the mathematical basis of the X pattern later. Although each layer line looks like a row of reflections, it is actually continuous intensity. This would be apparent if the pattern were plotted at higher overall intensity. The layer lines are numbered with integers from the equator ($l = 0$). Because of symmetry, the first lines above and below the equator are labeled $l = 1$, and so forth.

Compare helix (a) with (b), in which the helix has the same radius, but a longer *pitch P* (peak-to-peak distance). Note that the layer lines for (b) are more closely spaced. The layer-line spacing is inversely proportional to the helix pitch. The relationship is identical to that in crystals between lattice spacing and unit-cell dimensions [(Eq. (4.10)]. So precise measurement of layer-line spacing allows determination of helix pitch.

Helix (c) has the same pitch as (a) but a larger radius r. Notice that the X pattern in the transform is narrower than that of helices (a) and (b), which have the same radius. The angle formed by the branches of the X with the meridian, shown as the angle δ, is determined by the helix radius. But it appears at first that the relationship between δ and helix radius must be more complex, because angles δ for helices (a) and (b), which have the same radius, appear to be different. However, we define δ as the angle whose tangent is the distance w from the meridian to the center of the first intensity peak divided by the layer-line number l. You can see that the distance w at the tenth layer line is the same in (a) and (b). Defined in this way, and measured at relatively large layer-line numbers (beyond which the tangent of an angle is simply proportional to the length of the side opposite), δ is inversely proportional to the helix radius. Because the layer-line spacing is the same in (a) and (c), it is clear that an increase in radius decreases δ, and that we can determine the helix radius from δ.

Helix (d) has the same pitch and radius as helix (a), but is a helix of discrete objects or "repeats," like a polymeric chain of repeating subunits. The transform appears at first to be far more complex, but it is actually only slightly more so. It is merely a series of X patterns distributed along the meridian of the transform. To picture how multiple X patterns arise from a helix of discrete objects, imagine that the helix beginning with arbitrarily chosen object number 1 produces the X at the center of the transform. Then imagine

Figure 9.2 Helices and their Fourier transforms. (a) Simple, continuous helix. The first intensity peaks from the centers of each row form a distinctive X pattern. (b) Helix with longer pitch than (a) gives smaller spacing between layer lines. (c) Helix with larger radius than (a) gives narrower X pattern. (d) Helix of same dimensions as (a) but composed of discrete objects gives X patterns repeated along the meridian.

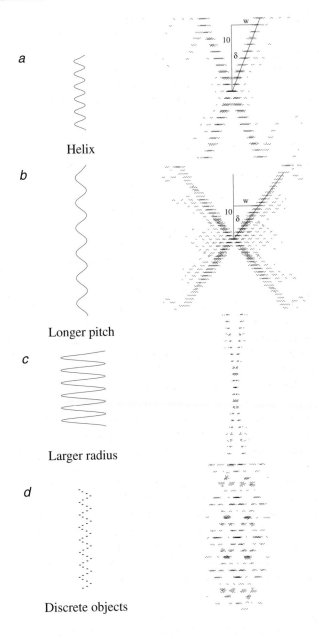

a

Helix

b

Longer pitch

c

Larger radius

d

Discrete objects

that the same helix beginning with object number 2 produces an X of its own. The distance along the meridian between centers of the two Xs is inversely proportional [(Eq. 4.10) again] to the distance between successive discrete objects in the helix. So careful measurement of the distance between successive meridional "reflections" (remember that the intensities on a layer line are actually continuous) allows determination of the distance between successive subunits of the helix. In a polymeric helix like that of a protein or nucleic acid, this parameter is called the rise-per-residue or p. Dividing the pitch P by the rise-per-residue p gives the number of residues per helical turn.

In addition, for the simplest type of discrete helix, in which there is an integral number of residues per turn of the helix, this integer P/p is the same as number l of the layer line on which the first meridional intensity peak occurs. Note that helix d contains exactly six residues per turn, and that the first meridional intensity peak above or below the center of the pattern occurs on layer line $l = 6$. To review, the layer-line spacing Z is proportional to $1/P$, and the distance from the origin to the first meridional reflection is proportional to $1/p$.

If the number of residues per turn is not integral, then the diffraction pattern is much more complex. For example, a protein alpha helix has 3.6 residues per turn, which means 18 residues in five turns. The diffraction pattern for a discrete helix of simple objects (say, points) with these dimensions has layer lines at all spacings $Z = (18m + 5\alpha)/5P$, where m and α are integers, and the diffraction pattern will repeat every 18 layer lines. But of course, protein alpha helix does not contain 3.6 simple points per turn, but instead 3.6 complex groups of atoms per turn. Combined with the rapid drop-off of diffraction intensity at higher diffraction angles, this makes for diffraction patterns that are too complex for detailed analysis.

Now let's look briefly at just enough of the mathematics of fiber diffraction to explain the origin of the X patterns. Whereas each reflection in the diffraction pattern of a crystal is described by a Fourier series of sine and cosine waves, each *layer line* in the diffraction pattern of a noncrystalline fiber is described by one or more *Bessel functions*, graphs that look like sine or cosine waves that damp out as they travel away from the origin (Fig. 9.3). Bessel functions appear when you apply the Fourier transform to helical objects. A Bessel function is of the form

$$J_\alpha(x) = \sum_{n=0}^{\infty} \frac{(-1)^n}{n!(n+\alpha)!} (x/2)^{(2n+\alpha)}. \tag{9.1}$$

The variable α is called the order of the function, and the values of n are integers. To plot the Bessel function of order zero, you plug in the values $\alpha = 0$ and $n = 0$ and then plot J as a function of x over some range $-x$ to $+x$. Next you would plug in $\alpha = 0$ and $n = 1$, plot again, and add the resulting curve to the one for which $n = 0$, just as curves were added together to give the Fourier series in Fig. 2.14. Continuing in this way, you would find that eventually, for large values of n, the new curves are very flat and do not change the previous sum. Bessel functions of orders zero, one, and two, for positive values of x, are shown in Fig. 9.3 a. Notice that as the order increases, the position of the first peak of the function occurs farther from the origin.

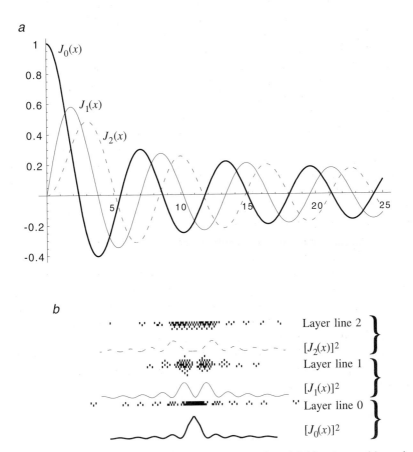

Figure 9.3 (a) Bessel functions J(x) of order $\alpha = 1, 2$, and 3 (showing positive values of x only). Note that, as the order increases, the distance to the first peak increases. (b) Enlargement of a few layer lines from Figure 9.2a, showing the correspondence between diffraction intensities and the squares of Bessel functions having the same order as the layer-line number.

Francis Crick showed in his doctoral dissertation that in the transform of a continuous helix, the intensity along a layer line is described by the square of the Bessel function whose order α equals the number l of the layer line, as shown in Fig. 9.3b, which is an enlargement of three layer lines from the diffraction pattern of Fig. 9.2a. Thus, the intensity of the central layer line, layer-line zero, varies according to $[(J_0(x)]^2$, which is the square of Eq. 9.1 with $\alpha = 0$. The intensity of the first line above (or below) center varies according to $[(J_1(x)]^2$, and so forth. This means that, for a helix, the first and largest peak of intensity lies farther out from the meridian on each successive layer line. The first peaks in a series of layer lines thus form the X pattern described earlier. The distance to the first peak in each layer line decreases as the helix radius increases, so thinner helices give wider X patterns.

For helices with a nonintegral number of residues per turn, the intensity functions, like the layer-line spacing, are also more complex, with two or more Bessel functions contributing to the intensities on each layer line. For the α helix, with 18 residues in five turns, the Bessel functions that contribute to layer line l are those for which α can be combined with some integral value of m (positive, negative or zero) to make $l = 18\alpha + 5m$ an integer. For example, for layer line $l = 0$, one solution to this equation is $\alpha = m = 0$, so $J_0(x)$ contributes to layer line 0. So also does $J_5(x)$, because $\alpha = 5$, $m = -18$ also gives $l = 0$. You can use $l = 18\alpha + 5m$ also to show that $J_2(x)$ ($m = -7$) and $J_7(x)$ ($m = -25$) contribute to layer line $l = 1$, but that $J_1(x)$ does not.

Probably the most famous fiber diffraction patterns are those of A-DNA and B-DNA obtained by Rosalind Franklin and shown in Fig. 9.4. Franklin's sample of A-DNA was microcrystalline, so its diffraction pattern (a) contains discrete reflections, many of them overlapping at the higher diffraction angles. The B-DNA was noncrystalline, so the intensities in its diffraction pattern (b) vary smoothly across each layer line. Considering the B-helix, the narrow spacing between layer lines is inversely proportional to its 34-Å pitch. The distance from the center to the strong meridional intensity near the edge of the pattern is inversely proportional to the 3.4-Å rise per subunit (a nucleotide pair, we now know). Dividing pitch by rise gives 10 subunits per helical turn, as implied by the strong meridional intensity at the tenth layer line. Finally, the angle of the X pattern implies a helix radius of 20 Å.

Francis Crick recognized that Frankin's data implied a helical structure for B-DNA. Using helical parameters deduced from the pattern, knowing the chemical composition of DNA and the structures of DNA's molecular components, and finally, using the tantalizing observation that certain pairs of bases occur in equimolar amounts, James Watson was able to build a feasible model of B-DNA.

Why could Watson and Crick not simply back-transform the diffraction data, get an electron-density map, and fit a model to it? As in any diffraction

a   *b*

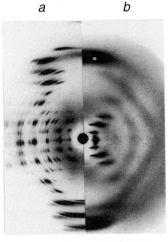

A-DNA   B-DNA

Figure 9.4   Fiber diffraction patterns from A-DNA (left half of figure) and B-DNA (right). A-DNA was microcrystalline and thus gave discrete, but overlapping, Bragg reflections. B-DNA was noncrystalline and thus gave continous variation in intensity along each layer line. Image kindly provided by Professor Kenneth Holmes.

experiment, we obtain intensities, but not phases. In addition, if we could somehow learn the phases, the back-transform would be the electron density averaged around the molecular axis parallel to the fiber axis, which would show only how electron density varies with distance from the center of the helix. Instead of map interpretation, structure determination by fiber diffraction usually entails inferring the dimensions of chains from the diffraction pattern and then building models using the known composition and the inferred dimensions, constrained by what is stereochemically allowable. As Watson's and Crick's success showed, for a helical substance, being able to deduce from diffraction the pitch, radius, and number of residues per helical turn puts some very strong constraints on a model.

The Fourier transform does, however, provide a powerful means of testing proposed models. With a feasible model in hand, researchers compute its transform and compare it to the pattern obtained by diffraction. If the fit is not perfect, they adjust the model, and again compare its transform with the X-ray pattern. They repeat this process until a model reproduces in detail the experimental diffraction pattern. Thus you can see why a very successful practitioner in this field, who may prefer to remain nameless, said, "Fiber diffraction is not what you'd do if you had a choice." Sometimes it is simply the only way to get structural information from diffraction.

For an example of structure analysis by fiber diffraction, see H Wang, H. and G. Stubbs, Structure determination of cucumber green mottle mosaic virus by X-ray fiber diffraction. Significance for the evolution of tobamoviruses, *J. Mol. Biol.* **239,** 371–384, 1994.

III. Diffraction by amorphous materials (scattering)

With fibers, the diffraction pattern, and hence any structural information contained therein, is averaged cylindrically about the molecular axis parallel to the fiber axis. This means that the transform of the diffraction pattern (computation of which would require phases) is an electron-density function showing only how electron density varies with distance from the molecular axis. For a helix of points aligned with the fiber axis, there would be a single peak of density at the radius of the helix. For a polyalanine helix (Fig. 9.5), there would be a density peak for the backbone, because atoms C, O, and CA are all roughly the same distance from the center of the helix (inner circle), and a second peak for the beta carbons, which are farther from the helix axis (outer circle). This averaging greatly reduces the structural information that can be inferred from the diffraction pattern, but it does imply that distance information is present, despite rotational averaging.

Imagine now a sample, such as a powder or solution, in which all the molecules are randomly oriented. Diffraction by such amorphous samples is usually called *scattering*. The diffraction pattern is averaged in all directions, spherically, because the X-ray beam encounters all possible orientations of the molecules in the sample. But the diffraction pattern still contains information about how electron density varies with distance from the center of the molecules that make up the sample. Obviously, for complex molecules, this information would be singularly uninformative. But if the molecule under study contains only a few atoms, or only a few that dominate diffraction (like metal atoms in a protein), then it may be possible to extract useful distance information from the way that scattered X-ray intensity varies with the angle of scattering from the incident beam of radiation. As usual, when we try to extract information from intensity measurements, we work without knowledge of phases.

I have shown that, in simple systems, Patterson functions can give us valuable clues about distances, even when we know nothing about phases (see Chapter 6, Section III.C). Diffraction from the randomly oriented molecules in a solution or powder would give a spherically averaged diffraction pattern, from which we can compute a spherically averaged Patterson map. Is this map interpretable?

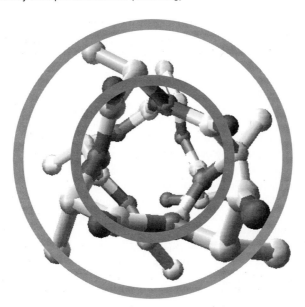

Figure 9.5 Polyalanine *a* helix, viewed down the helix axis. An electron-density map averaged around this axis would merely show two circular peaks of electron density, one at the distance of the backbone atoms (inner circle) and another at the distance of the alpha carbons (outer circle).

As in the case of heavy-atom derivatives, we can interpret a Patterson map if there are just a few atoms or a few very strong diffractors. Consider a linear molecule containing only three atoms (Fig. 9.6). We can see what to expect in a Patterson function computed from diffraction data on this molecule by constructing a Patterson function for the known structure. First, construct a Patterson function for the structure shown in Fig. 9.6*a*, using the procedure described in Fig. 6.10. The result is (*b*). Spherical averaging of (*b*) gives a set of spherical shells of intensity, with cross-section shown in (*c*). This cross section contains information about distances between the atoms in (*a*). The radius of the first circle is the bond length r, and the radius of the second circle is the length of the molecule ($2r$). A plot of the magnitude of the Patterson function as a function of distance from the origin, called a radial Patterson function, will contain peaks that correspond to vectors between atoms. Furthermore, the intensity of each peak will depend on how many vectors of that length are present. In the molecule of Fig. 9.6*a*, there are four vectors of length r and two vectors of length $2r$, so the Patterson peak at r is stronger than the one at $2r$. You can see that, for a small molecule, the radial Patterson function computed from scattering intensities may contain enough information to determine distances between atoms. For larger numbers of atoms, the radial Patterson function would contain peaks corresponding to all the interatomic distances.

The radial Patterson function of an amorphous (powder or dilute solution) sample of a protein contains an enormous number of peaks. But imagine a protein containing a cluster of one or two metal ions surrounded by sulfur atoms. These atoms may dominate the powder diffraction data, and the strongest peaks in the radial Patterson function may reveal the distances among the metal ions and sulfur atoms. Remember that we obtain distance information but no geometry because diffraction is spherically averaged and all directional information is lost. Sometimes powder or solution diffraction can be used to extract distance information relating to a cluster of the heavier atoms in a protein. These distances put constraints on models of the cluster. Researchers can compare spherically averaged back-transforms of plausible models with the experimental diffraction data to guide improvements in the model, as in fiber diffraction. In some cases, model building and comparison of model back-transforms with data allows indentification of ligand atoms and estimation of bond distances.

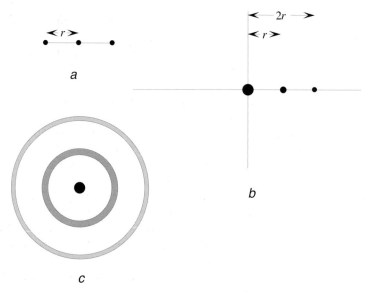

Figure 9.6   Radial Patterson function. (*a*) Linear triatomic molecule. (*b*) Patterson function constructed from (*a*). For construction procedure, see Chapter 6, Section III.C and Figure 6.10. (*c*) Patterson function (*b*) averaged by rotation, the expected result of calculating a Patterson function from diffraction intensities measured from an amorphous sample.

At very small diffraction angles, additional information can be obtained about the size and shape of a molecule. A detailed treatment of this method, called *low-angle scattering*, is beyond the scope of this book, but it can be shown that, for very small angles, the variation in scattering intensity is related to the *radius of gyration*, R_G, of the molecule under study. The radius of gyration is defined as the root-mean-square average of the distance of all scattering elements from the center of mass of the molecule. For two proteins having the same molecular mass, the one with the larger radius of gyration is the more extended or less spherical one. Combined knowledge of the molecular mass and the radius of gyration of a molecule allows an estimate of its shape.

The precise relationship between scattering intensity and radius of gyration is

$$\langle I(\theta) \rangle = n_e^2 (1 - 16\pi^2 R_G^2 \sin^2 \theta / 3\lambda^2). \tag{9.2}$$

$\langle I(\theta) \rangle$ is the intensity of scattering at angle θ, n_e is the number of electrons in the molecule, and λ is the X-ray wavelength. Equation (9.2) implies that a graph of radiation intensity versus $\sin^2 \theta$ has a slope that is directly proportional to the square of R_G. Note also that such a graph can be extrapolated to $\theta = 0$, where the second term in parentheses disappears, and $\langle I(\theta) \rangle$ is equal to the square of n_e, or roughly, to the square of the molecular mass. So for molecules about which very little is known, measurement of scattering intensity at low angles provides estimates of both molecular mass and shape.

This kind of information about mass, shape, and distance can be obtained on amorphous samples not only from X-ray scattering but also from scattering by other forms of radiation, including light, and as I will discuss in Section IV, neutrons. The choice of radiation depends on the size of the objects under study.

The variable-wavelength X rays available at synchrotron sources give researchers an additional, powerful way to obtain precise distance information from amorphous samples. At wavelengths near the absorption edge of a metal atom (Chapter 4, Section III.B), there is rapid oscillation of X-ray absorption as a function of wavelength. This oscillation results from interference between diffraction from the absorbing atom and that of its neighbors. The Fourier transform of this oscillation, in comparison with transforms calculated from plausible models, can reveal information on the number, types, and distances of the neighboring atoms. Distance information in favorable cases can be much more precise than atomic distances determined by X-ray crystallography. Thus this information can anticipate or add useful detail to crystallographic models. The measurement of absorbance as a function of wavelength is, of course, a form of spectroscopy. In the instance described here, it is called *X-ray absorption spectroscopy*, or XAS. Fourier analysis of near-edge X-ray absorption is called *extended X-ray absorption fine structure*, or EXAFS.

For an example of structure analysis by X-ray scattering, see D. I. Svergun, S. Richard, M. H. Koch, Z. Sayers, S. Kuprin, and G. Zaccai, Protein hydration in solution: experimental observation by X-ray and neutron scattering, *Proc. Natl. Acad. Sci. USA* **95,** 2267–2272, 1998.

IV. Neutron diffraction

The description of diffraction or scattering as a Fourier transform applies to all forms of energy that have wave character, including not only electromagnetic energy like X-rays and light but also subatomic particles, including neutrons and electrons, which have wavelengths as a result of their motion. The de Broglie equation (9.3) gives the wavelength λ of a particle of mass m moving at velocity v:

$$\lambda = h/mv, \tag{9.3}$$

where h is Planck's constant. We can use the de Broglie wavelength to describe diffraction of particles by matter. In this section, I will describe single-crystal neutron diffraction and neutron scattering by macromolecules, emphasizing the type of information obtainable. In the next section, I will apply these ideas to electron crystallography.

Recall that X rays are diffracted by the electrons that surround atoms, and that images obtained from X-ray diffraction show the surface of the electron clouds that surround molecules. Recall also that the X-ray diffracting power of elements in a sample increases with increasing atomic number. Neutrons are diffracted by nuclei, not by electrons. Thus a density map computed from neutron diffraction data is not an electon-density map, but instead a map of nuclear mass distribution, a "nucleon-density map" of the molecule (nucleons are the protons and neutrons in atomic nuclei).

The neutron crystallography experiment is much like X-ray crystallography (see Figs. 2.5 and 2.11). A crystal is held in a collimated beam of neutrons. Diffracted beams of neutrons are detected in a diffraction pattern that is a reciprocal-lattice sampling of intensities, but as usual not phases, of the Fourier transform of the average object in the crystal. Structure determination is, in principle, similar to that in X-ray crystallography, involving estimating the phases, back-transforming a set of structure factors composed of experimental intensities and estimated phases, improving the phases, and refining the structure.

There are two common ways to produce a beam of neutrons. One is steady-state nuclear fission in a reactor, which produces a continuous output of neutrons, some of which sustain fission, while the excess are recovered as a usable neutron beam. The second type is a pulsed source in which a cluster of protons or other charged particles from a linear accelerator are injected into a synchrotron, condensed into a tighter cluster or pulse, and allowed to strike a target of metal, such as tungsten. The high-energy particles drive neutrons from the target nuclei in a process called *spallation*. Neutrons from both fission and spallation carry too much energy for use in diffraction, so they are slowed or cooled ("thermalized") by passage through heavy water (D_2O) at 300 K, producing neutrons with De Broglie wavelengths ranging from 1 to 2 Å. This wavelength is in the same range as X rays used in crystallography.

Thermal neutrons then enter a collimator followed by a monochromator, which selects a narrow range of wavelengths to emerge and strike the sample. Monochromators are single crystals of graphite, zinc, or copper. They act like diffraction gratings to direct neutrons of different energies (and hence, wavelengths) in different directions, as a prism does with light.

The collimated, monochromatic neutron beam is delivered to the sample on a diffractometer, and diffraction is detected by an area detector (Chapter 4, Section III.C). The most common type is a multiwire area detector that uses helium-3 as the active gas, according to this reaction:

$$^3\text{He} + n \rightarrow \,^1\text{H} + \,^3\text{H} + \text{energy.}$$

A second type of image-plate detector employs gadolinium oxide, which absorbs a neutron and emits a gamma ray, which in turn exposes the image plate. Image plates have higher spatial resolution but lower efficiency than multiwire detectors.

The great advantage of neutron diffraction is that small nuclei like hydrogen are readily observed. By comparison with carbon and larger elements, hydrogen is a very weak X-ray diffractor and is typically not observable in electron-density maps of proteins. But hydrogen and its isotope deuterium (^2H or D) diffract neutrons very efficiently in comparison with larger elements.

The concept of *scattering length* is used to compare diffracting power of elements. The scattering length b (do not confuse it with temperature factor B) is the amplitude of scattering at an angle of zero degrees to the incident beam and is the absolute measure of scattering power of an atom. Measured diffraction intensities are proportional to b^2, which is why X-ray diffraction from a single heavy atom can be easily detected above the diffraction of all the small atoms in a massive macromolecule. Table 9.1 gives neutron and X-ray scattering lengths for various elements and allows us to compare the scattering power of elements in neutron and X-ray diffraction.

Table 9.1

X-ray and Neutron Scattering Lengths of Various Elements

Element	X rays $b \times 10^{13}$ (cm)	Neutrons $b \times 10^{13}$ (cm)
H	3.8	−3.74
D	2.8	6.67
C	16.9	6.65
N	19.7	9.40
O	22.5	5.80
P	42.3	5.10
S	45	2.85
Mn	70	−3.60
Fe	73	9.51
Pt	220	9.5

Data from C. R. Cantor and P. R. Schimmel. (1980). *Bio-physical Chemistry, Part II: Techniques for the Study of Biological Structure and Function.* W. H. Freeman and Company, San Francisco, p. 830.

Note that, even though H and D are weak X-ray diffractors compared to other elements in biomolecules (they have so few electrons around them), they are comparable to other elements when it comes to neutron diffraction. So hydrogen will be a prominent feature in density maps from neutron diffraction. Note also that the sign of b is negative for H. This means that H diffracts with a phase that is opposite to that of other elements. As mentioned before, measured diffraction intensities are related to b^2, so the negative sign of b has no observable effect on diffraction patterns. But in density maps, H gives negative density, which makes it stand out. In addition, the large magnitude of the difference between scattering lengths for H and D allows some powerful ways to use D as a label in diffraction or scattering experiments.

First, let us consider experiments analogous to single-crystal X-ray crystallography. In most electron-density maps of macromolecules, we cannot observe hydrogens. Thus we cannot distinguish the amide nitrogen and oxygen in glutamine and asparagine side chains. Neither can we determine the locations of hydrogens on histidine side chains, whose pK_a values allow both protonated and unprotonated forms at physiological pH. And we cannot detect critical hydrogens involved in possible hydrogen bonding with ligands like cofactors and transition-state analogs. Some hydrogens, including those on hydoxyl O and amide N, can exchange with hydrogens of the solvent if they are exposed, and if they are not involved in tight hydrogen bonds. Distinguishing

H from D would mean being able to determine which hydrogens are exchangeable. Neutron diffraction, it would seem, gives us a way to answer these questions.

We can collect diffraction intensities from macromolecular crystals, but can we phase them and thus obtain maps that include clear images of hydrogen atoms? How about heavy atoms for phasing? According to Table 9.1, there is no such thing as a heavy atom! In other words, no nucleus diffracts so strongly that we can detect it above all others in a Patterson map. That rules out MIR, SIR with anomalous dispersion, and MAD as possible phasing methods. But if we are trying to find hydrogens in a structure known, or even partially known, from X-ray work, we have a source of starting phases in hand.

A crystal has the same reciprocal lattice for all types of diffraction, because the construction of the reciprocal lattice (Chapter 4, Section I.D) does not depend in any way upon the type of radiation involved. So if we know the positions of all or most of the nonhydrogen atoms from X-ray structure determination, we can compute their contributions to the neutron-diffraction phases. These contributions depend only on atomic positions in the unit cell, and like reciprocal lattice positions, they are independent of the type and wavelength of radiation. We start phasing by assigning the final phases computed from the X-ray model of nonhydrogen atoms to the reflections obtained by neutron diffraction. From a Fourier synthesis combining neutron-diffraction intensities and X-ray phases, we compute the first map. Then we proceed as usual, alternating cycles of examining the map, building a model (in this case, adding hydrogens whose images we see in the map), back-transforming the model to get better phases, and so forth. In essence, this is isomorphous molecular replacement (Chapter 6, Section V), using a hydrogen-free model that is identical to the fully hydrogenated model we seek.

Neutron diffraction has been used to detect critical hydrogen bonds in protein-ligand complexes. For example, the presence of a hydrogen bond between dioxygen and the distal histidine of myoglobin is known because of neutron diffraction studies. In highly refined neutron-diffraction models, hydrogen positions are determined as precisely as positions of other atoms. In the best cases, not only can hydrogens in hydrogen bonds be seen, but it is even possible to tell whether hydrogen-containing groups like methyl or hydroxyl can rotate. If methyl groups are conformationally locked, density maps will show three distinct peaks of density for the three hydogens of a methyl group, whereas if dynamic rotation is possible, or if alternative static conformations occur on different molecules, then we observe a smooth donut of density around the methyl carbon. For hydroxyls, it is possible to determine the conformation angle H-O-C-H in hydrogen-bonding and nonhydrogen-bonding situations. In the former, neutron diffraction sometimes reveals unexpected eclipsed conformations of the hydroxyl. Finally, the structure of networks of water molecules on the surface of macromolecules have been revealed by neutron diffraction. Being able to image the hydrogens means learning the orientation of each water molecule and the exact pattern of hydrogen bonding. In electron-density maps from X-ray work, we usually learn only the location of the oxygen atom of water molecules.

Because of the difference in signs of b for H and D, it is easy to distinguish them in density maps. This makes it possible to detect exchangeable hydrogens in proteins by comparing the density maps from crystals in H_2O and in D_2O. Amide hydrogens exhange with solvent hydrogens only if the amide group is exposed to solvent and is not involved in tight hydrogen bonding to other atoms. Because X-ray data are taken over a long period of time compared to rapid proton exchange, proton exchange rates can usually only be assigned to three categories: no exchange, slow exchange (10–60% during the time of data collection), and fast exchange (more than 60%).

Turning to scattering by amorphous samples, and to studies at lower resolution than crystallography, the negative scattering length for H makes possible interesting and useful scattering experiments on macromolecular complexes in solution. For example, neutron scattering in solution can be used to measure distances between the various protein components of very large macromolecular complexes like ribosomes and viruses. Understanding these methods requires bringing up a point that I have been able to avoid until now. Specifically, all forms of scattering depend on contrast between the scatterer and its surroundings. Even in single-crystal X-ray crystallography, we get our first leg up on phases by finding the molecular boundary, in essence distinguishing protein from solvent. This is possible because of the contrast in density between ordered protein and disordered water. If the protein and water had exactly the same scattering power, we could not find this crucial boundary.

The importance of contrast to scattering is analogous to the importance of refractive index to refraction, such as the curving of light beams when they enter water. Light travels in a straight line through a pure liquid, but changes direction abruptly when it crosses a boundary into a new medium having a different refractive index. If two liquids have identical refractive indices, the boundary between them does not bend a light ray. Analogously, if X rays or neutrons pass from one medium (say, solvent) into another (say, protein), scattering occurs only if the average scattering lengths of the media differ.

Because H and D have scattering lengths of different sign, it is possible to make mixtures of H_2O and D_2O with average scattering lengths over a wide range. In addition, by preparing proteins containing varying amounts of D substituted for H (by growing protein-producing cells in H_2O/D_2O mixtures), researchers can prepare proteins of variable average scattering length. A protein that scatters identically with the solvent is invisible to scattering. Imagine then reconstituting a bacterial ribosome, which is a complex of

3 RNA molecules and about 50 proteins, from partially D-labeled proteins and RNAs whose scattering power matches the H_2O/D_2O mixture in which they are dissolved. This mixture will not scatter neutrons. But if two of the components are unlabeled, only those two will scatter. At low resolution, it is as if the two proteins constitute one molecule made of two large atoms. The variation in intensity of neutron scattering with scattering angle will reveal the distance between the two proteins, just as our hypothetical radial Patterson function reveals interatomic distances in Fig. 9.6. Repeating the experiment with different unlabeled pairs of proteins gives enough interprotein distances to direct the building of a three-dimensional map of protein locations.

Finally, low-angle neutron scattering can provide information about the shape of these molecules within the ribosome, which may not be the same as their shape when free in solution.

Neutron diffraction experiments are generally more difficult than those involving X-rays. There are fewer neutron sources worldwide. Available beam intensities, along with the low neutron-scattering lengths of all elements, translate into long exposure times. But for most macromolecules, neutron diffraction is the only source of detailed information about hydrogen locations and the exact orientation of hydrogen-carrying atoms.

For an example of neutron diffraction applied to a crystallographic problem, see D. Pignol, J. Hermoso, B. Kerfelec, I. Crenon, C. Chapus, and J. C. Fontecilla-Camps, The lipase/colipase complex is activated by a micelle: Neutron crystallographic evidence, *Chem. Phys. Lipids* **93**, 123–129, 1998.

V. Electron diffraction

Electrons, like X rays and neutrons, are scattered strongly by matter and thus are potentially useful in structure determination. Electron microscopy (EM) is the most widely known means of using electrons as structural probes. Scanning EM gives an image of the sample surface, which is usually coated with a thin layer of metal. Sample preparation techniques for scanning EM are not compatible with obtaining images of molecules at atomic resolution. Transmission EM produces a projection of a very thin sample or section onto a viewing screen. In the most familiar electron micrographs of cells and organelles, the sample is stained with metals to outline the surfaces of membranes and large multimolecular assemblies like ribosomes. Unfortunately, staining results in distortion of the sample that is unacceptable at high resolution.

The electrons produced by transmission electron microscopes, whose design is analogous to light microscopes, have de Broglie wavelengths of less than 0.1 Å, so they are potentially quite precise probes of molecular structure. Unlike X rays and neutrons, electrons can be focused (by electric fields rather than glass lenses) to produce an image, although direct images of objects in transmission EM do not approach molecular resolution. However, electron microscopes can be used to collect electron-diffraction patterns from two-dimensional arrays of molecules, such as closely packed arrays of membrane proteins in a lipid layer. Analysis of diffraction by such two-dimensional "crystals" is called *electron crystallography.*

Among the main difficulties with electron crystallography are (1) sample damage from the electron beam (a 0.1-Å wave carries a lot of energy), (2) low contrast between the solvent and the object under study, and (3) weak diffraction from the necessarily very thin arrays that can be studied by this method. Despite these obstacles, cryoscopic methods (Chapter 3, Section V) and image processing techniques have made electron crystallography a powerful probe of macromolecular structure, especially for membrane proteins, many of which resist crystallization.

Transmission electron microscopy is analogous to light microscopy, with visible light replaced by a beam of electrons produced by a heated metal filament, and glass lenses replaced by electromagnetic coils to focus the beam. An image of the sample is projected onto a fluorescent screen or, for a permanent record, onto film or a CCD detector (Chapter 4, Section III.C). Alternatively, an image of the sample's diffraction pattern can be projected onto the detectors.

To see how we can observe either an image or a diffraction pattern, look again at Fig. 2.1, which illustrates the action of a simple lens, such as the objective (lower) lens of a microscope. Recall that the lens produces an image at I of an object O placed outside the front focal point F of the lens. In a light microscope, an eyepiece (upper) lens is positioned so as to magnify the image at I for viewing. If we move the eyepiece to get an image of what lies at the back focal plane F', we see instead the diffraction pattern of the sample. Analogously, in EM, we can adjust the focusing power of the lenses to project an image of the back focal plane onto the viewing screen, and thus see the sample's diffraction pattern. If the image is nonperiodic (say, a section of a cell) the diffraction pattern is continuous, as in Plates 8 and 9. If the sample is a periodic two-dimensional array, the diffraction pattern is sampled at reciprocal-lattice points, just as in X-ray diffraction by crystals (Plate 3*e*). The Fourier transform of this pattern is an image of the average object in the periodic array. As with all diffraction, producing such an image requires knowing both diffraction intensities and phases.

The possibility of viewing both the image and the diffraction pattern is unique to electron crystallography and, in favorable cases, can allow phases to be determined directly. For an ordered array (a 2-D crystal) of proteins in a lipid membrane, the direct image is, even at the highest magnification, a featureless gray field. But phase estimates can be obtained from this singularly uninteresting image in the following manner. The image is digitized to pixels at high resolution, producing a two-dimensional table of *image*-intensity values (*not* diffraction intensities). This table is then Fourier transformed. Computing the FT of a table of values is the same process as used to produce the images in Plates 3, 8, and 9, in which the "samples" are in fact square arrays of pixels with different numerical values. If the gray EM image is actually a periodic array, the result of the FT on the table of pixel values is the diffraction pattern of the average object in the array, sampled at reciprocal-lattice points. But because this transform is computed from "observed" objects, it includes both intensities and phases. The intensities are not as accurate as those that we can measure directly in the diffraction plane of the EM, but the phases are often accurate enough to serve as first estimates in a refinement process.

Single-crystal X-ray crystallography requires measuring diffraction intensities at many closely spaced crystal orientations, so as to measure most of the unique reflections in the reciprocal lattice. The Fourier transform (with correct phases) of reflections from a single orientation gives only a projection of the unit cell in a plane perpendicular to the beam. The transform of the full data set gives a three-dimensional image. Electron microscopes allow samples to be tilted, which for diffraction is analogous to rotating the crystal. The EM sample can be tilted about 75° at most, which may or may not allow a sufficiently large enough portion of the reciprocal lattice to be sampled. It is not unusual for EM data sets to be missing data in parts of reciprocal space that cannot be brought into contact with the sphere of reflection (Figs. 4.10, 4.11, and 4.24) by tilting. Most common is for a cone of reflections to be missing, with the result that maps computed from these data do not have uniform resolution in all directions.

Armed with sets of data measured at different sample tilt angles, each set including (1) intensities measured at the diffraction plane and (2) phases from the Fourier transform of direct images, the crystallographer can compute a map of the unit cell. Recall that X rays are scattered by electrons around atoms, producing a map of electron density, and neutrons are scattered by nuclei, producing a map of mass or nucleon density. Electrons, in turn, are scattered by electrostatic interactions, producing maps of electrostatic potential, sometimes called electron-potential maps, or just potential maps. Maps from electron diffraction, like all other types, are interpreted by building molecular models to fit them and refined by using partial models as phasing models.

One of the special features of electron crystallography is the possibility of detecting charge on specific atoms or functional groups. The probing electrons interact very strongly with negative charge, with the result that negatively charged atoms have negative scattering factors at low resolution. (Recall that the negative scattering length for H in neutron diffraction can make H particularly easy to detect in neutron-density maps.) At high resolution, the negative sign of the scattering factor weakens the signal of negatively charged groups, but not usually enough to be obvious. However, a comparison between maps computed with and without the low-angle data can reveal charged groups. If the two maps are practically identical around a possibly charged group like a carboxyl, then the group is probably neutral. If, however, the map computed without low-angle data shows a stronger density for the functional group than does the map computed from all data, then the functional group probably carries a negative charge. For example, functional groups in proton pumps like rhodopsin are apparently involved in transferring protons across a membrane by way of a channel through the protein. Determining the ionization state of functional groups in the channel is essential to proposing pumping mechanisms.

Another group of powerful structural methods involving EM are collectively called *image enhancement.* These methods do not involve diffraction directly, but they take advantage of the averaging power of Fourier transforms to produce high-resolution direct images. Image enhancement can greatly improve the resolution of supramolecular complexes like ribosomes, viruses, and multienzyme complexes that can be seen individually, but at low resolution, as direct EM images.

Imagine a direct EM image of, say, virus particles, all strewn across the viewing field. Particles lay before you in all orientations, and the images show very little detail. In image enhancement, we digitize the individual images and sort them into images that appear to share the same orientation. Then we can compute the Fourier transforms of the images. These transforms should be similar in appearance if indeed the images in a set share the same orientation. Next, we align the transforms, add them together, and back-transform. The result is an averaged image, in which details common to the component images are enhanced, and random differences, a form of "noise," are reduced or eliminated. This process is then repeated for sets of images at different orientations, the transforms are all combined into a three-dimensional set, and the back-transform gives a three-dimensional image.

Recall that the Fourier transform of a crystal diffraction pattern gives an image of the average molecule in the crystal. Because any averaging process tends to eliminate random variations and enhance features common to all components, the image is much more highly resolved than if it were derived from a single molecule. In like manner, the final back-transform of Fouriers from many EM images is an image of the average particle in the direct EM field, and the resolution can be greatly enhanced in comparison to any single particle image. In favorable cases, enhanced EM images can be used as phasing models in crystallographic structure determination by molecular replacement, where the particles or even components of the particles can be crystallized.

For an example of electron diffraction in structure determination of a membrane protein, see Y. Kimura, D. G. Vassylyev, A. Miyazawa, A. Kidera, M. Matsushima, K. Mitsuoka, K. Murata, T. Hirai, and Y. Fujiyoshi, Surface of bacteriorhodopsin revealed by high-resolution electron crystallography, *Nature* **389,** 206–211, 1997.

VI. Laue diffraction and time-resolved crystallography

Now I return to X-ray diffraction to describe probably the oldest type of diffraction experiment, but one whose stock has soared with the advent of synchrotron radiation and powerful computer techniques for the analysis of complex diffraction data. The method, Laue diffraction, is already realizing its promise as a means to determine the structures of short-lived reaction intermediates. This method is sometimes called *time-resolved crystallography,* implying an attempt to take snapshots of a chemical reaction or physical change in progress.

Laue crystallography entails radiating a crystal with a powerful *polychromatic* beam of X rays, whose wavelengths range over a two- to threefold range, for example, 0.5–1.5 Å. The resulting diffraction patterns are more complex than those obtained from monochromatic X rays, but they sample a much larger portion of the reciprocal lattice from a single crystal orientation. To see why this is so, look again at Fig. 4.10, which demonstrates in reciprocal space the conditions that satisfy Bragg's law, and shows the resulting directions of diffracted rays when the incident radiation is monochromatic. Recall that Bragg's law is satisfied when rotation of the crystal (and with it, the reciprocal lattice) brings point P (in Fig. 4.10a) or P' [in (b)] onto the surface of the sphere of reflection. The results are diffracted rays R and R'. Also recall that the sphere of reflection, whose radius is the reciprocal of the X-ray wavelength λ, passes through the origin of the reciprocal lattice, and its diameter is coincident with the X-ray beam.

Figure 9.7 extends the geometric construction of Fig. 4.10 to show the result of diffraction when the X-ray beam provides a continuum of wavelengths. Instead of one sphere of reflection, there are an infinite number, covering the gray region of the figure, with radii ranging from $1/\lambda_{max}$ to $1/\lambda_{min}$, where λ_{max} and λ_{min} are the maximum and minimum wavelengths of radiation in the beam. In the figure, spheres of reflection corresponding to λ_{max} and λ_{min} are shown, along with two others that lie in between. The fours points lying on the incident beam X are the centers of the four spheres of reflection. Each of the four spheres is shown passing through one reciprocal-lattice point (some pass through others also), producing a diffracted ray (R_1 through R_4).

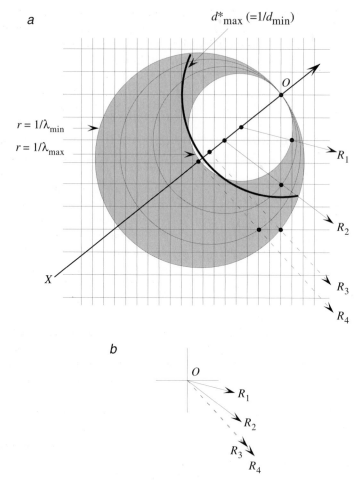

Figure 9.7 Geometric construction for Laue diffraction in reciprocal space.

Note, however, that because there are an infinite number of spheres of reflection, *every* reciprocal-lattice point within the gray region lies on the surface of some sphere of reflection, and thus for every such point Bragg's law is satisfied, giving rise to a diffracted ray. Of course, any crystal has its diffraction limit, which depends on its quality. The heavy arc labeled d^*_{max} represents the resolution limit of the crystal in this illustration. Diffraction corresponding to reciprocal-lattice points outside this arc (for example, R_3 and R_4), even though they satisfy Bragg's law, will not be detected simply because this crystal does not diffract out to those high angles. Because the detector is at the geometric equivalent of an infinite distance from the crystal, we can picture all rays as emerging from a single point, such as the origin. To construct the Laue pattern from this diagram, we move all the diffracted rays to the origin, as shown for rays R_1 through R_4 in Fig. 9.7b.

Note also that because of the tapering shape of the gray area near the origin, the amount of available data drops off at low angles. One limitation of Laue diffraction is the scarcity of reflections at small angles. An added complication of Laue diffraction is that some rays pass through more than one point. When this occurs, the measured intensity of the ray is the sum of the intensities of both reflections. Finally, at higher resolution, many reflections overlap; for example, note that rays R_3 and R_4 are superimposed when displaced to the origin.

You can see from Fig. 9.8 that a Laue diffraction pattern is much more complex than a diffraction pattern from monochromatic X rays. But modern software can index Laue patterns and thus allow accurate measurement of many diffraction intensities from a single brief pulse of X rays through a still crystal. If the crystal has high symmetry and is oriented properly, a full data set can in theory be collected in a single brief X-ray exposure. In practice, this approach usually does not provide sufficiently accurate intensities because the data lack the redundancy necessary for high accuracy. Multiple exposures at multiple orientations are the rule.

The unique advantage of the Laue method is that data can be collected rapidly enough to give a freeze-frame picture of the crystal's contents. Typical X-ray data are averaged over the time of data collection, which can be hours, days, or even months, and over the sometimes large number of crystals required to obtain a complete data set. Laue data has been collected with X-ray pulses shorter than 200 picoseconds. Such short time periods for data collection are comparable to half-times for chemical reactions, especially those involving macromolecules, such as enzymatic catalysis. This raises the possibility of determining the structures of reaction intermediates.

Recall that a crystallographic structure is the average of the structures of all diffracting molecules, so structures of intermediates can be determined when all molecules in the crystal react in unison or exist in an intermediate state simultaneously. Thus the crystallographer must devise some way to trigger the reaction simultaneously throughout the crystal. Good candidate reactions include those triggered by light. Such processes can be initiated throughout a crystal by a laser pulse. Some reactions of this type are reversible, so the reaction can be run repeatedly with the same crystal.

A strategy for studying enzymatic catalysis entails introducing into enzyme crystals a "caged" substrate—a derivative of the substrate that is prevented from reacting by the presence of a light-labile protective group. If the caged substrate binds at the active site, then the stage is set to trigger the enzymatic reaction by a light pulse that frees the substrate from the protective group,

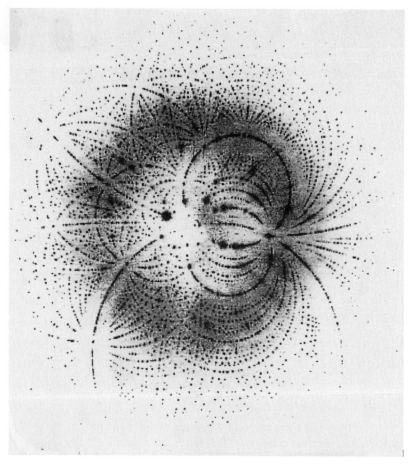

Figure 9.8   A typical Laue image.

allowing the enzyme to act. Running this type of reaction repeatedly may mean replacing the crystal, or maybe just introducing a fresh supply of caged substrate to diffuse into the crystal for the next run.

Other possibilities are reactions that can be triggered by sudden changes in temperature or pressure. Reactions that are slow in comparison to diffusion rates in the crystal may be triggered by simply adding substrate to the mother liquor surrounding the crystal and allowing the substrate to diffuse in. If there is a long-lived enzyme-intermediate state, then during some interval after substrate introduction, a large fraction of molecules in the crystal will exist in this state, and a pulse of radiation can reveal its structure.

In practice, multiple Laue images are usually necessary to give full coverage of reciprocal space with the required redundancy. Here is a hypothetical strategy for time-resolved crystallography of a reversible, light-triggered process. From knowledge of the kinetics of the reaction under study, determine convenient reaction conditions (for example, length and wavelength of triggering pulse and temperature) and times after initiation of reaction that would reveal intermediate states. These times must be long compared to the shortest X-ray pulses that will do the job of data collection. From knowledge of crystal symmetry, determine the number of orientations and exposures that will produce adequate data for structure determination. Then set the crystal to its first orientation, trigger the reaction, wait until the first data-collection time, pulse with X rays, and collect Laue data. Allow the crystal to equilibrate, which means both letting the reaction come back to equilibrium and allowing the crystal to cool, since the triggering pulse and the X-ray pulse heat the crystal. Then again trigger the reaction, wait until the second data-collection time, pulse with X rays, and collect data. Repeat until you have data from this crystal orientation at all the desired data-collection times. Move the crystal to its next orientation, and repeat the process. The result is full data sets collected at each of several time intervals during which the reaction was occurring.

For an example of Laue diffraction applied to time-resolved crystallography, see V. Srajer, T. Teng, T. Ursby, C. Pradervand, Z. Ren, S. Adachi, W. Schildkamp, D. Bourgeois, M. Wulff, and K. Moffat, Photolysis of the carbon monoxide complex of myoglobin: Nanosecond time-resolved crystallography, *Science* **274**, 1726–1729, 1996.

VII. Conclusion

The same geometric and mathematical principles lie at the root of all types of diffraction experiments, whether the samples are powders, solutions, fibers, or crystals, and whether the experiments involve electromagnetic radiation (X rays, visible light) or subatomic particles (electrons, neutrons). My aim in this chapter was to show the common ground shared by all of these probes of molecular structure. Note in particular how the methods complement each other and can be used in conjunction with each other to produce more inclusive models of macromolecules. For example, phases from X-ray work can serve as starting phase estimates for neutron work, and the resulting accurate coordinates of hydrogen positions can then be added to the X-ray model. As another example, direct images obtained from EM work (sometimes after image enhancement) can sometimes be used as molecular-replacement models for X-ray crystallography.

As a user of macromolecular models, you are faced with judging whether each model really supports the insights it appears to offer. The principles presented in Chapter 7, on how to judge the quality of models, apply to models obtained from all types of diffraction experiments. But today's structural databases also contain a growing number of models obtained by methods other than diffraction. In the next chapter, I will describe the origin of the major types of non-diffraction models and provide some guidance on how to use them wisely.

10 Other Kinds of Macromolecular Models

I. Introduction

When you go looking for models of macromolecules that interest you, crystallographic models are not the only type you will find. As of the end of 1998, about 15% of the models in the Protein Data Bank are derived from NMR spectroscopy of macromolecules in solution. Most all of these models are proteins of less than 150 residues, but the number of larger models is sure to increase. Also proliferating rapidly are *homology models*, which are built by computer algorithms that work on the assumption that proteins of homologous sequence have similar three-dimensional structures. Massive databases of homology models are now under construction, with the goal of providing homology models for all known protein sequences that are homologous to structures determined by crystallography or NMR. Although this effort is just beginning, the number of homology models available in one database, the SWISS-MODEL Repository (see the CCMC Home Page), already dwarfs the Protein Data Bank. Furthermore, it is now possible for you to determine protein structures by homology modeling on your personal computer, as I will describe in Chapter 11. Finally, there are various means of producing theoretical models, for example, based on attempts to simulate folding of proteins. As with crystallographic models, other types of models vary widely in quality and reliability. The user of these models is faced with deciding whether the quality of the model allows confidence in the apparent implications of a structure.

In this chapter, I will provide brief descriptions of how protein structures are determined by NMR and by homology modeling. In addition, I will provide some guidance on judging the quality of noncrystallographic models, primarily by drawing analogies to criteria of model quality in crystallography. Recall that in all protein structure determination, the goal is to determine the conformation of a molecule whose chemical composition (amino-acid content and sequence) is known.

II. NMR models

A. Introduction

Models of proteins in solution can be derived from NMR spectroscopy. In brief, the process entails collecting highly detailed spectra; assigning spectral peaks (*resonances*) to all residues by chemical shift and by decoupling experiments; deriving distance restraints from couplings, which reveal local conformations and also pinpoint pairs of atoms that are distant from each other in sequence, but near each other because of the way the protein is folded; and computing a chemically, stereochemically, and energetically feasible model that complies with all distance restraints. The power of NMR spectroscopy has grown immensely with the development of pulse-Fourier transform instruments, which allow rapid data collection; more powerful magnets, which increase spectral resolution; and sophisticated, computer-controlled pulse sequences, which spread data over multiple dimensions to reveal through-bond and through-space couplings.

In this section, I provide a simplified physical picture of pulse NMR spectroscopy, including a simple conceptual model to help you understand multidimensional NMR. Then I briefly discuss the problems of assigning resonances and determining distance restraints for molecules as large and complex as proteins, and the methods for deriving a structure from this information. Finally, I discuss the contents of coordinate files from NMR structure determination and provide some hints on judging the quality of models.

B. Principles

I assume that you are conversant with basic principles of ^{1}H or proton NMR spectroscopy as applied to small molecules. In particular, I assume that you understand the concepts of chemical shift (δ) and spin-spin coupling, classical continuous-wave methods of obtaining NMR spectra, and decoupling experiments to determine pairs of coupled nuclei. If these ideas are unfamiliar to you, you may wish to review NMR spectroscopy in an introductory organic chemistry textbook before reading further.

Chemical shift and coupling

Many atomic nuclei, notably ^{1}H, ^{13}C, ^{15}N, ^{19}F and ^{31}P, have net nuclear spins as a result of the magnetic moments of their component protons and neutrons. These spins cause the nuclei to behave like tiny magnets, and as a result, to adopt preferred orientations in a magnetic field. A nucleus having a spin quantum number of $1/2$ (for example, ^{1}H) can adopt one of two orientations in a magnetic field, aligned either with or against the field. Nuclei aligned with the field are slightly lower in energy, so at equilibrium, there are slightly more nuclei (about 1 in 10,000) in the lower energy state. The orientations of nuclear spins can be altered by pulses of electromagnetic radiation in the radio-frequency (RF) range.

Nuclei in different chemical environments absorb different frequencies of energy. This allows specific nuclei to be detected by their characteristic absorption energy. This energy can be expressed as an RF frequency (in hertz), but because the energy depends on the strength of the magnetic field, it is expressed as a frequency difference between that of the nucleus in question and a standard nucleus (like hydrogen in tetramethylsilane, a common standard for ^{1}H NMR) divided by the strength of the field. The result, called the *chemical shift* δ and expressed in parts per million (ppm), is independent of field strength but varies informatively with the type of nucleus and its immediate molecular environment. Figure 10.1 shows the ^{1}H-NMR spectrum of human thioredoxin,[1] a small protein of 105 amino-acid residues. (In humans, thioredoxin plays a role in activating certain transcriptional and translational regulators by a dithiol/disulfide redox mechanism. More familiar to students of biochemistry is thioredoxin's role in green plants: mediating light activation of enzymes in the Calvin cycle of photosynthesis.) The spectrum is labeled with ranges of δ for various types of hydrogen atoms in proteins.

[1]J. D. Forman-Kay, G. M. Clore, P. T. Wingfield, and A. M. Gronenborn, High-resolution three-dimensional structure of reduced recombinant human thioredoxin in solution, *Biochemistry* **30**, 2685, 1991 (PDB files 2trx and 3trx).

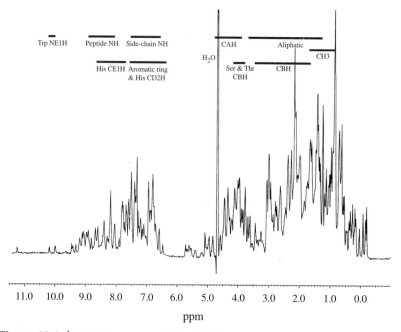

Figure 10.1 ^{1}H-NMR spectrum of thioredoxin, reduced form. Labels show chemical-shift values typical of various hydrogen types in protein chains having random coil conformation. Some signals lie outside these ranges because of specific interactions not present in random coils. Atom labels are as found in PDB coordinate files. Spectrum generously provided by Professor John M. Louis.

In addition to having characteristic chemical shifts, nuclear spins interact magnetically with each other by a process called *spin-spin coupling*. Coupling distributes or splits the absorption signals of nuclei about their characteristic absorption frequency, usually in a distinctive pattern that depends on the number of equivalent nuclei that are coupled. The spacing between signals produced by splitting is called the coupling constant J, which is expressed in hertz because its magnitude does not depend on field strength. Because nuclei must be within a few bonds of each other to couple, this effect can be used to determine which nuclei are neighbors in the molecule under study.

Absorption and relaxation

You can view the nuclear spins in a sample as precessing about an axis, designated z, aligned with the magnetic field H, as shown in Fig. 10.2, which shows

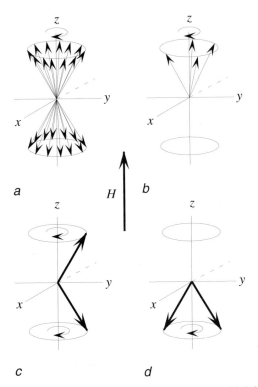

Figure 10.2 (*a*) Nuclear spins precess around the *z*-axis, which is parallel to the applied field *H*. Each spin precesses at a rate that depends on its RF absorption frequency. (*b*) Excess spins in the lower energy state. At equilibrium, there are slightly more spins aligned with *H* than against *H*. (*c*) Immediately after a 90° pulse, equal numbers of spins are aligned with and against *H*, and initially all spins lie in the *yz*-plane, giving a net magnetic vector on the *y*-axis. As time passes after the pulse, spins precess about *z* and spread out, due to their different precession frequencies, ultimately returning to the state shown in *a*. The result is a decay in signal intensity. (*d*) Immediately after a 180° pulse, or two successive 90° pulses, excess spins shown in (*b*) are inverted, and lie in the *yz*-plane as shown. The net magnetization vector in the *xy*-plane has a magnitude of zero, and thus induces no signal in the receiver coils.

the spins for a large number of equivalent nuclei in (*a*), and only the slight equilibrium excess of spins at the lower energy in (*b*). The slight excess of spins aligned with the field means that there is a net magnetization vector pointing in the positive direction along *z* at equilibrium. Spins precess at their characteristic RF absorption frequency, but the phases of precession are random. In other words, equivalent, independent spins precess at the same rate, but their positions are randomly distributed about the *z*-axis.

In Fig. 10.3, a highly simplified sketch of an NMR instrument shows the orientation of the x-, y-, and z-axes with repect to the magnets that apply the field H, the sample, and wire coils that transmit and detect radio-frequency signals. The receiver coils, which encircle the x-axis, detect RF radiation in the xy-plane only. A single nuclear spin precessing alone around the z-axis has a rotating component in the xy-plane, and transmits RF energy at its characteristic absorption frequency to the receiver coils, in theory revealing its chemical shift. But when many equivalent spins in a real sample are precessing about the z-axis with random phase, the xy components of their magnetic vectors cancel each other out, the net magnetic vector in the xy plane has a magnitude of zero, and no RF signals are detected.

Detection of RF radiation from the sample requires that a net magnetization vector be moved into the xy-plane. This is accomplished by a wide-band (multifrequency) pulse of RF energy having just the right intensity to tip the net magnetization vector into the xy-plane. This pulse is applied along x, and it equalizes the number of spins in the higher and lower energy states, while aligning them onto the yz-plane (Fig. 10.2c). A pulse of this intensity is call a 90° pulse. The result is that a net magnetization vector appears in the xy-plane and begins to rotate, generating a detectable RF signal. Because nuclei of different chemical shifts precess at different frequencies, their net magnetization vectors in the xy-plane rotate at different frequencies, and the resulting RF signal contains the characteristic absorption frequencies of all nuclei in the sample, from which chemical shifts can be derived.

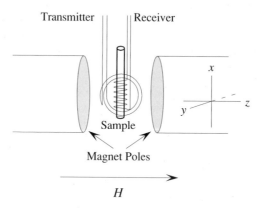

Figure 10.3   Diagram of an NMR experiment. The sample lies between the poles of a powerful magnet, and is spun rapidly around its long (x) axis in order to compensate for any unevenness (inhomogeneity) of the magnetic field. Radio frequency receiver coils form a helix around the x-axis, and transmitter coils spiral around the y-axis.

As shown in Fig. 10.2c, the 90° pulse equalizes the number of nuclei aligned with and against the applied field along z, giving a net magnetization vector along z of zero. This means that the system of spins is no longer at equilibrium. It will return to equilibrium as a result of spins losing energy to their surroundings, a process called *spin-lattice relaxation*, in which the term *lattice* simply refers to the surroundings of the nuclei. This relaxation is a first-order process whose rate constant I will call R_L (L for lattice). The inverse of R_L is a time constant I will call T_L, the spin-lattice relaxation time constant. T_L is also called the *longitudinal* relaxation time constant, because relaxation occurs along the z-axis, parallel to the magnetic field. (T_L is traditionally called T_1, but I adopt a more descriptive symbol to reduce confusion with other symbols in the following discussion.)

After a 90° pulse, another relaxation process also occurs. Although all spins have the same phase just after the pulse (aligned along y), the phases of identical nuclei spread out due to exchanges of energy that result from coupling with other spins. The result is that the net magnetization vector in the xy-plane for each distinct set of nuclei diminishes in magnitude, as individual spin magnetizations move into orientations that cancel each other. Furthermore, the RF frequencies of chemically distinct nuclei disappear from the overall signal at different rates, depending on their coupling constants, which reflect the strength of coupling or, in other words, how effectively the nuclei exchange energy. Phases of pairs of nuclei coupled to each other spread out at the same rate because their spin energies are simply being exchanged. The rate constant for this process, which is called *spin-spin relaxation*, is traditionally called R_2, but I will call it R_S (S for spin). Its inverse, the time constant T_S, is the spin-spin relaxation time constant. It is also called the *transverse* relaxation time constant because the relaxation process is perpendicular or transverse to the applied magnetic field H. Each set of chemically identically nuclei has a characteristic T_S. (Often, and confusingly, the rate constants R are loosely called rates, and time constants T are simply called relaxation times.)

So after a 90° RF pulse applied along x tips the net magnetization vector onto the y-axis, an RF signal appears at the detector because there is net magnetization rotating in the xy-plane. This signal is a composite of the frequencies of all the precessing nuclei, each precessing at its characteristic RF absorption frequency. In addition, this signal is strong at first but decays because of spin-lattice and spin-spin relaxation. Signals for different nuclei diminish at different rates that depend on their spin-lattice and spin-spin relaxation rates. This complex signal is called a *free-induction decay* or *FID: free* because it is free of influence from the applied RF field, which is turned off after the pulse; *induction* because the magnetic spins induce the signal in the receiver coil; and *decay* because the signal decays to an equilibrium value of zero. A typical hydrogen NMR FID signal decays in about 300 ms, as shown in Fig. 10.4.

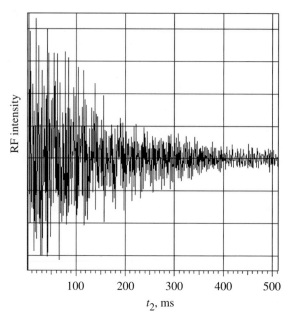

Figure 10.4 Free induction decay signal, which appears in the receiver coils after a 90° pulse. The FID is a time-domain spectrum, showing RF intensity as a function of time t_2. It is a composite of the RF absorption frequencies of all nuclei in the sample. The Fourier transform decomposes an FID into its component frequencies, giving a spectrum like that shown in Fig. 10.1. Figure generously provided by Professor John M. Louis.

One-dimensional NMR

How do we extract the chemical shifts of all nuclei in the sample from the free-induction decay signal? The answer is our old friend the Fourier transform. The FID is called a *time-domain* signal because it is a plot of the oscillating and decaying RF intensity *versus* time, as shown in Fig. 10.4 (the time axis is conventionally labeled t_2, for reasons you will see shortly). Fourier transforming the FID produces a *frequency-domain* spectrum, a plot of RF intensity versus the frequencies present in the FID signal, with the frequency axis labeled v_2 for frequency or F_2 for chemical shift, as shown in Fig. 10.1. So the Fourier transform decomposes the FID into its component frequencies, revealing the chemical shifts of the nuclei in the sample.

NMR spectroscopists collect this type of classical or one-dimensional (1-D) NMR spectra on modern FT-NMR instruments by applying a 90° pulse and collecting the FID signal that is induced in the receiving coil. To make a stronger signal, they collect FIDs repeatedly and add them together. Real signals appear at the same place in all the FIDs and add up to a large sum. On the other hand, random variation or "noise" appears in different places in different FIDs, so their signals cancel each other. The summed FIDs thus have a high signal-to-noise ratio, and FT produces a clean spectrum with well-resolved chemical shifts. From here on, I may not always mention that pulse/FID collection sequences are repeated to improve the signal, but you can assume that all sequences I describe are carried out repeatedly, commonly 64 times, for this purpose.

Two-dimensional NMR

Couplings between nuclei influence the rate at which their characteristic frequencies diminish in the FID. If we could measure, in addition to the frequencies themselves, their rates of disappearance, we could determine which pairs of nuclei are coupled, because the signals of coupled pairs fade from the FID at the same rate. This is the basis of *two-dimensional* (2-D) NMR, which detects not only the chemical shifts of nuclei but also their couplings. The 2-D NMR employs computer controlled and timed pulses that allow experimenters to monitor the progress of relaxation for different sets of nuclei.

Figure 10.5 illustrates, with a system of four nuclei, the principles of 2-D NMR in its simplest form, when we want to assign pairs of ^1H nuclei that are spin-spin coupled through a small number of bonds, such as hydrogens on adjacent carbon atoms. You may find this figure daunting at first, but careful study as you read the following description will reward you will a clearer picture of just what you are seeing when you look at a 2-D NMR spectrum.

The 2-D experiment is much more complex than obtaining a conventional "one-dimensional" spectrum. The experimenter programs a sequence of pulses, delays, and data (FID) collections. In each sequence, the program directs a 90° pulse (the "preparation" pulse), followed by a delay or "evolution" time t_1, and then a second 90° "mixing" pulse, followed by "detection" or data collection (all repeated 64 times to enhance the signal). In successive preparation/evolution/mixing/detection (PEMD) sequences, the evolution time t_1 is increased in equal steps. In a typical experiment of this type, t_1 might be incremented 512 times in 150-μs steps from 0 μs up to a maximum delay of about 77 ms, giving a full data set of 512 signal-enhanced FIDs, five of which are shown with their FTs in Fig. 10.5a. Each FID is a plot of RF signal intensity versus time t_2, and its FT shows intensity as a function of frequency F_2. In each of the FTs (except the first one, in which the signals are too weak—see Fig. 10.1d to see why), you can see the RF signals of the four nuclei in the sample.

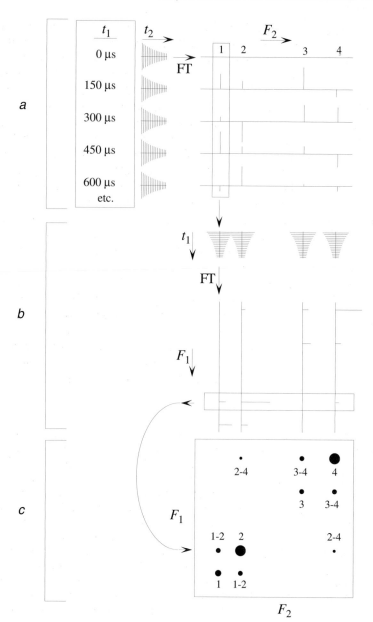

What is the purpose of the second or mixing pulse? Recall that the first pulse tips the precessing magnetic vectors toward the y-axis and aligns their phases, thus putting rotating net magnetic vectors in the xy-plane and an FID signal into the detector. But in 2-D NMR, we do not record this FID. Instead, we wait for a specified evolution time t_1 (between 0 and 77 ms in the example described here), and then pulse again. During this interval—the evolution time—magnetic vectors spread out in the xy-plane, and the RF signal diminishes in intensity. The second pulse tips any magnetization that currently has a component in the yz-plane (as a result of vector spreading) back onto the xy-plane, and aligns the spins on the y-axis. Then relaxation occurs again, and this time we record the FID. In the first sequence, with $t_1 = 0$, essentially no relaxation occurs between pulses, no magnetization enters the yz-plane, and after the second pulse, the FID contains weak or no signals. As t_1 is increased, more relaxation occurs between the first and second pulses, magnetization representative of the precession state of each nucleus at time t_1 is present in the yz-plane, and it is detected after the second pulse.

The FIDs are Fourier transformed to produce 512 frequency spectra, each from a different time t_1, as shown on the right in Fig. 10.5a. Just as in 1-D NMR, each spectrum gives the intensities of RF signals as a function of frequency F_2 but, in this case, recorded from all nuclei after relaxation for a time t_1. To assign couplings, we are interested in finding out which pairs of RF frequencies F_2 vary in intensity at the same rate. This can be accomplished by plotting the signal at each frequency F_2 versus time t_1. The data for such a

Figure 10.5 Two-dimensional NMR for a hypothetical system of four nuclei. (a) FIDs are collected after each of a series of sequences [90° pulse/t_1 delay/90° pulse], with t_1 varied. Each FID decays over time t_2. The Fourier transform of an FID gives the RF intensities of each frequency F_2 in the FID at time t_1. Each signal occurs at the chemical shift F_2 of a nucleus in the sample. Chemical shifts of the nuclei are labeled 1 through 4. Intensity at a single frequency F_2 evolves over time t_1, as shown in the narrow vertical rectangle. (b) A plot of this variation is like an FID that decays over time t_1 [top of (b)]. Fourier transforms of t_1-FIDs reveal the frequencies F_1 present in the signal. Coupled nuclei have frequencies F_1 in common, as shown for nuclei 1, 2, and 4 in the narrow horizontal rectangle. They share decay frequencies because they are coupled, which means they are exchanging energy with each other. (c) Contour plot presenting the information of FTs in (b). Peaks on the diagonal correspond to the 1-D NMR spectrum, with F_1 and F_2 both corresponding to chemical shifts. Rows of peaks off the diagonal indicate sets of nuclei that have frequencies F_1 in common due to coupling. The third horizontal row of signals in (c) corresponds to the information in the rectangle in (b). In this example, the couplings are 1 to 2, 2 to 1 and 4, 3 to 4, and 4 to 2 and 3. From *Principles of Biophysical Chemistry* by Van Holde, copyright 1998; adapted by permission of Prentice-Hall, Inc., Upper Saddle River, NJ.

plot for spin #1 is shown within the narrow vertical rectangle in Fig. 10.5a, and the plot itself is shown by the FID symbol at the top left in Fig. 10.5b. As implied by the symbol for this plot, at a frequency F_2 at which an RF signal occurs, a plot of its intensity versus t_1 is a frequency-domain spectrum, actually a pattern of interference among all frequencies present. This pattern is mathematically like an FID, and for convenience, it is usually called an FID. Recall that FIDs are composed of one or more component frequencies. If a nucleus is coupled to more than one other nucleus, its FID taken over time t_1 will contain frequencies (corresponding to the spreading of vectors in the xy-plane) corresponding to all of its couplings. To determine these frequencies, we once again use the Fourier transform. The FT of this FID will reveal the different frequencies for each coupling. This FT is a plot of RF signal intensity versus phase-spreading frequencies or spin-spin-relaxation frequencies F_1, shown for the four F_2 signals in the lower part of Fig. 10.5b. The frequencies F_1 in each spectrum are relaxation frequencies for a given nucleus. The intensities in these spectra give the strength of coupling, or to put it another way, they give the amount of correlation between nuclei whose F_1 FIDs contain the same frequency.

A 2-D NMR spectrum is a plot of F_1 versus F_2, that is, it is our set of 512 F_1 spectra laid side by side at a spacing corresponding to the frequency F_2 at which each F_1 FID was taken, as in Fig. 10.5b. Looking along horizontals across the F_1 spectra in (b), we see the frequencies F_2 of sets of nuclei that relaxed at the same rate. For example, the narrow horizontal rectangle in Fig. 10.5b shows that spins 1, 2, and 4 share a common relaxation frequency. Thus we conclude that these nuclei are coupled to each other and are on neighboring atoms. The 2-D spectrum is usually presented as shown in Fig. 10.5c. The position and intensity of a spot in this spectrum corresponds to the position and intensity of a signal in (b).

The off-diagonal or "cross" peaks have finer details, not shown here, that correspond to the simple and familiar, but information-rich, J-splitting patterns, such as doublets, triplets, and quartets, seen in conventional ^{1}H-NMR spectra of small molecules. These patterns tell us how many chemically equivalent nuclei are involved in these couplings. Of course, each signal correlates most strongly with itself over time t_1, so the strongest signals lie on the diagonal. The spectrum on the diagonal corresponds to the one-dimensional NMR spectrum.

Interpretation of the 2-D spectrum in Fig. 10.5c is simple. The four spins give strong signals on the diagonal (numbered 1 through 4). All off-diagonal signals indicate couplings. Each such signal is aligned horizontally along F_2 and vertically along F_1 with signals on the diagonal that correspond to the two nuclei, or sets of nuclei, that are coupled. The signals labeled 2–4 denote coupling between nuclei 2 and 4. This type of 2-D NMR spectroscopy, which reveals spin-spin or J couplings by exhibiting correlations between spin-spin relaxation times of nuclei, is called correlation spectroscopy or COSY. In a sense, it spreads the information of the one-dimensional spectrum into two dimensions, keeping coupled signals together by virtue of their correlated or shared relaxation times. By far the most widely useful nucleus for NMR COSY study of proteins is ^1H. Obtaining ^1H spectra (rather than ^{13}C or ^{15}N spectra, for example) simply means carrying out the NMR experiments described here in the relatively narrow range of RF frequencies over which all ^1H atoms absorb energy in a magnetic field.

Nuclear Overhauser effect

The COSY spectra reveal through-bond couplings because these couplings are involved in the relaxation process revealed by this PEMD sequence of operations: 90° pulse/t_1 delay/90° pulse/data collection. In a sense, this pulse sequence stamps the t_1 coupling information about relaxation rates onto the t_2 RF absorption signals. The second set of Fourier transforms extracts this information. Other pulse sequences can stamp other kinds of information onto the t_2 signal.

In particular, there is a second form of coupling between nuclear dipoles that occurs through space, rather than through bonds. This interatomic interaction is called the nuclear Overhauser effect (NOE), and the interaction can either weaken or strengthen RF absorption signals. Nuclei coupled to each other by this effect will show the effect to the same extent, just as J-coupled nuclei share common spin-spin relaxation rates. Appropriate pulse sequences can impress NOE information onto the t_2 absorption signals, giving 2-D NMR spectra in which the off-diagonal peaks represent NOE couplings, rather than J couplings, and thus reveal pairs of nuclei that are near each other in space, regardless of whether they are near each other through bonds. This form of 2-D NMR is called NOESY. Though the off-diagonal signals represent different kinds of coupling, NOESY spectra look just like COSY spectra.

A partial NOESY spectrum of human thioredoxin is shown in Fig. 10.6. This 2-D spectrum shows the region from $\delta = 6.5$ to $\delta = 10$. First, compare this spectrum with the 1-D spectrum in Fig. 10.1. Recall that the signals on the diagonal of a 2-D spectrum are identical to the 1-D spectrum, but they are in the form of a contour map, sort of like looking at the 1-D spectrum from above, or down its intensity axis. For example, notice the two small peaks near $\delta = 10$ on the 1-D spectrum, and near $F_1 = F_2 = 10$ on the 2-D spectrum. Notice that there are off-diagonal peaks aligned horizontally and vertically with these peaks. These off-diagonal signals align with other diagonal peaks corresponding to nuclei NOE-coupled to these nuclei. Several coupling assignments that were used to define distance restraints are indicated by pairs of lines—one horizontal, one vertical—on the spectrum. The pairs converge on an off-diagonal peak and diverge to the signals of coupled nuclei on the diagonal.

NMR in higher dimensions

For a large protein, even spectra spread into two dimensions by J or NOE coupling are dauntingly complex. Further simplification of spectra can be accomplished by three- and higher-dimensional NMR, in which the information of

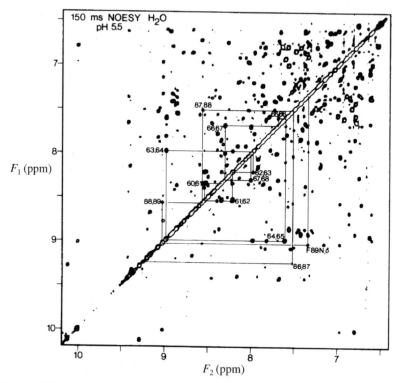

Figure 10.6   NOESY spectrum of thioredoxin in the region $\delta = 6.5$ to $\delta = 10$. Pairs of lines—one horizontal, one vertical—converge at off-diagonal peaks indicating NOE couplings and diverge to the signals on the diagonal for the coupled nuclei. Assigning resonances on the diagonal to specific hydrogens in the protein requires greater resolution and simplification of the spectrum than shown here. The off-diagonal peak labeled "F89N,δ" indicates a specific NOESY interaction that is described and shown in Plate 14. Spectrum generously provided by Professor John M. Louis.

a complex 2-D spectrum is separated onto sets of spectra that can be pictured as stacked planes of 2-D spectra, as in Fig. 10.7. One technique separates the 2-D spectrum into separate planes, each corresponding to a specific chemical shift of a different NMR-active nucleus, such as ^{13}C or ^{15}N, to which the hydrogens are bonded. This method, called ^{13}C- or ^{15}N-editing, requires that the protein be heavily labeled with these isotopes, which can be accomplished by obtaining the protein from cells grown with only ^{13}C- and ^{15}N-containing nutrients. In ^{13}C-edited COSY or NOESY spectra, each plane contains cross-peaks only for those hydrogens attached to ^{13}C atoms having the chemical shift corresponding to that plane. Because you can picture these planes as stacked along a ^{13}C chemical-shift axis perpendicular to F_1 and F_2 (as illustrated in Fig. 10.7b), this type of spectroscopy is referred to as three-dimensional NMR. Further separation of information on these planes onto additional planes, say of ^{15}N chemical shifts (Fig. 10.7c), is the basis of so-called 4-D and higher dimensional NMR. The dimensions referred to are not really spatial dimensions, but are simply subsets of the data observed in 2-D spectra.

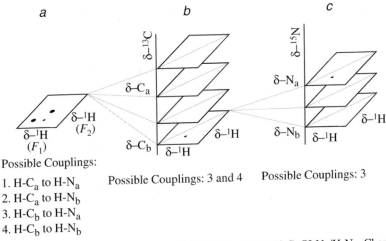

Possible Couplings:

1. H-C$_a$ to H-N$_a$ Possible Couplings: 3 and 4   Possible Couplings: 3
2. H-C$_a$ to H-N$_b$
3. H-C$_b$ to H-N$_a$
4. H-C$_b$ to H-N$_b$

Figure 10.7   3-D and 4-D NMR. (a) 2- D NMR of H-C$_a$/H-C$_b$/H-N$_a$/H-N$_b$. Chemical shifts of H-C$_a$ and H-C$_b$ are identical, as are those of H-N$_a$ and H-N$_b$. so only two peaks appear on the diagonal. One cross-peak could arise from four possible couplings, listed on the figure. (b) Spectrum of (a) spread onto planes, each at a different ^{13}C chemical shift δ. The δ-^{13}C axis represents the third NMR dimension. Thin lines represent the position of the diagonal on each plane. The cross-peak occurs only on the plane at δ-^{13}C$_b$, revealing that coupling involves H-C$_b$ and eliminating two possible couplings. (c) Spectrum at in (b), spread onto planes, each at a different ^{15}N chemical shift δ. The δ-^{15}N axis represents the fourth NMR dimension. The cross-peak occurs only on the plane at δ-^{15}N$_a$, revealing that coupling involves H-N$_a$. Thus the coupling indicated by the-cross peak in (a) is H-C$_b$ to H-N$_a$.

Figure 10.7 shows how 3-D and higher dimensional NMR can simplify spectra and reveal the details of couplings. Consider a hypothetical example of four hydrogens—two bonded to nitrogen, N_a-H and N_b-H, and two bonded to carbon, C_a-H and C_b-H. The two hydrogens of each pair have the same chemical shift, so the 1-D NMR spectrum shows two peaks, shown on the diagonal in Fig. 10.7a. In addition, there is one cross-peak, indicating that one H-C is coupled to one H-N. There are four possibilities for this coupling, listed in (a). If we separate these spectra, as shown in (b), into planes that show only cross-peaks for protons attached to ^{13}C atoms having a narrow range of chemical shifts, the cross-peak occurs only on the plane corresponding to C_b. This means that C_b carries one of the coupled hydrogens and reduces the possible couplings to those listed as 3 and 4 in (a). If we again separate the spectra, as in (c), into planes based on ^{15}N shifts, we find that the cross-peak occurs only on the N_a plane, eliminating all but possibility 3, coupling of C_b-H to N_a-H. Spreading the information onto additional planes in this manner does not really produce any new information. Instead it divides the information into subsets that may be easier to interpret.

C. Assigning resonances

The previous section describes methods that can provide an enormous amount of chemical-shift and coupling information about a protein. Recall that our goal is to determine the protein's conformation. We hope that we can use couplings to decide which pairs of hydrogens are neighbors, and that this information will restrict our model's conformations to one or a few similar possibilities. But before we can use the couplings, we must assign all the resonances in the 1-D spectrum to specific protons on specific residues in the sequence. This is usually the most laborious task in NMR structure determination, and I will provide only a brief sketch of it here.

The 1-D NMR spectrum of thioredoxin in Fig. 10.1 shows the range of chemical shifts δ for various proton types found in a protein. Because there are only about 20 different types of residues present, and because residues have many proton types in common (most have one proton on amide N, one on C_α, two on C_β, and so forth), the spectrum is composed of several very crowded regions. Obviously, we could never assign specific signals to specific residues without greatly expanding or simplifying the spectrum. A variety of through-bond correlation and NOE experiments allow us to resolve all these signals. Given the necessary resolution, we are faced first with identifying individual residue types, say, distinguishing alanines from valines.

The cross-peaks in through-bond spectra guide us in this task, because each residue type has a characteristic set of splittings that makes it identifiable. This allows us to determine which of the many C_α, C_β, and C_γ signals, each in its own characteristic region of the spectrum, belong to the same residue. Thus spin systems that identify specific residue types can be identified and grouped together. This is usually the first stage of analysis: identification of spin systems and thus of groups of signals that together signify individual residues. There is conformational information in these couplings also, because the precise values of J-coupling constants depend on the conformational angles of bonds between atoms carrying coupled atoms. So the splitting patterns reveal residue identities, and actual values of splitting constants put some restrictions on local conformations.

Next, of all those spins systems identified as, say, valine, how do we determine which valine is valine-35, and which is valine-128? The next part of the analysis is making this determination, for which we must examine connectivities between residues. You can imagine that we might step from atom to atom through the protein chain by finding successively coupled proton pairs. But because proton spin-spin couplings are usually observed through at most three bonds (for example, H-C-N-H), and because the carbonyl carbon of each residue carries no hydrogens, proton spin-spin couplings do not extend continuously through the chain. In between two main-chain carbonyl carbons, the spin-spin coupling systems are isolated from each other. But there are interresidue spin-spin couplings because the N-H of residue n can couple to the C_α-H of residue n-1.

More recently, connectivities between residues are determined using proteins labeled fully with ^{13}C and ^{15}N, by way of various three-dimensional double and triple resonance experiments, each of which reveals atoms at opposite ends of a series of one-bond couplings. In essence, a sequence of pulses transfers magnetization sequentially from atom to atom. Coupling sequences such as H-N-C_α, H-N-(CO)-C_α, H-N-(CO), H-C_α-(CO), and H-C_α-(CO)-N can be used to "walk" through the couplings in a protein chain, establishing neighboring residues.

Taking NOE connectivities into account brings more potential correlations to our aid. There is a high probability that at least one proton among the N-H, C_α-H, or C_β-H of residue n will be within NOE distance of the N-H of residue n+1. This means that NOE correlations can also help us determine which residues are adjacent. In the end, this kind of analysis yields a complete set of resonance assignments. In addition, it reveals much about backbone and side-chain conformations because knowing which protons interact by NOE greating restricts the number of possible conformations. These restrictions are a key to determining the conformation of the protein.

D. Determining conformation

Recall that our goal is to determine, in detail, the conformation of the protein. From the analysis described in the previous section, we know which signals in the spectra correspond to which residues in the sequence. In addition, the magnitudes of J couplings put some restraints on local conformational angles, both main-chain and side-chain, for each residue. What is more, the specific NOE couplings between adjacent residues also constrain local conformations to those that bring the coupled atoms to within NOE distance of each other (the range of distances detected by NOESY can be determined by the details of pulse timing during data collection).

Nuclear Overhauser effect correlations must be interpreted with care, because the NOE couples atoms through space, not through bonds. So an NOE coupling between protons on two residues may not mean that they are adjacent in sequence. It may instead mean that the protein fold brings the residues near to each other. Distinguishing NOE cross-peaks between sequential neighbors from those between residues distant in the sequence sets the stage for determining the conformation of the protein. Once we have assigned all neighboring-residue NOESY cross-peaks properly, then the remaining cross-peaks tell us which hydrogens of sequentially distant residues are interacting with each other through space. This adds many of the most powerful and informative entries to the list of distance restraints with which our final model must agree.

The end result of analyzing NMR spectra is a list of distance restraints. The list tells us which pairs of hydrogens are within specified distances of each other in space. This is really about all that we learn about the protein from NMR. But we already know lots more. We know its chemical structure in detail because we know the full sequence of amino-acid residues, as well as the full structures of any cofactors present. We know with considerable precision all of the covalent bond lengths and bond angles. We know that most single-bond dihedral angles will lie within a few degrees of staggered conformations. We know that amide single-bonds dihedral angles will be close to 180°. We know that main-chain conformational Φ angles Ψ and will lie in the allowed zones of the Ramachandran diagram. We know all these things about any protein having a known sequence. The question is, does all this add up to knowing the three-dimensional conformation of the protein?

Next we would like to build a model of the protein that fits all we know, including the distance restraints we learn from NMR. This is no trivial task. Much research has gone into developing computer algorithms for this kind of model building. One general procedure entails starting from a model of the protein having the known sequence of residues, and having standard bond lengths and angles but random conformational angles. This starting structure will, of course, be inconsistent with most of the distance and conformational restraints derived from NMR. The amount of inconsistency can be expressed as a numerical parameter that should decline in value as the model improves, in somewhat the same fashion as the R-factor decreases as a crystallographic model's agreement with diffraction data improves during refinement.

Starting from a random, high-temperature conformation, simulated annealing or some form of molecular dynamics is used to fold our model under the influence of simulated forces that maintain correct bond lengths and angles, provide weak versions of van der Waals repulsions, and draw the model toward allowed conformations, as well as toward satisfying the restraints derived from NMR. Electrostatic interactions and hydrogen-bonding are usually not simulated, in order to give larger weight to restraints based on experimental data; after all, we want to *discover* these interactions in the end, not build them into the model before the data have had their say. The simulated folding process entails satisfying restraints locally at first, and then gradually satisfying them over greater distances. At some points in the procedure, the model is subjected to higher-temperature dynamics, to shake it out of local minima of the consistency parameter that might prevent it from reaching a global best fit to the data. Near the end, the strength of the simulated van der Waals force is raised to a more realistic value, and the model's simulated temperature is slowly brought down to about 25° C.

Next the model is examined for serious van der Waals collisions, and for large deviations from even one distance or conformational restraint. Models that suffer from one or more such problems are judged not to have converged to a satisfactory final conformation. They are discarded. The entire simulated folding process is carried out repeatedly, each time from a different random starting conformation, until a number of models are found that are chemically realistic and consistent with all NMR-derived restraints. If some of these models differ markedly from others, the researchers may try to seek more distance restraints in the NMR spectra that will address specific differences, and repeat the process with additional restraints. It may be possible at various stages in this process to use the current models to resolve previous ambiguities in NOE assignments and to include them in further model building. When the group of models appears to contain the full range of structures that satisfy all restraints, this phase of structure determination is complete.

The result of the NMR structure determination is thus not a single model, but a group or *ensemble* of from half a dozen to over 30 models, all of which agree with the NMR data. The models can be superimposed on each other by least-squares fitting (Chapter 7, Section V.A, and Chapter 11, Section IV.A) and displayed as shown in Plate 14*a* for thioredoxin (PDB 4trx). In this display of 10 the 33 models obtained, it is readily apparent that all models agree with each other strongly in some regions, often in the interior, whereas there is more variation in other regions, usually on the surface.

Researchers interested in understanding this protein will immediately want to use the results of the NMR work as an aid to understanding the function of this protein. Which of these perhaps two or three dozen models should we use? For further studies, as well as for purposes of illustration, it is appealing to desire a single model, but one that will not let us forget that certain regions may be more constrained by data than others, and that certain regions may be different in different models within the ensemble. The usual procedure is to compute the average position for each atom in the model and to build a model of all atoms in their average position. This model may be unrealistic in many respects. For example, bond lengths and angles involving atoms in their averaged positions may not be the same as standard values. This averaged model is then subjected to restrained energy minimization, which in essence brings bond lengths and angles to standard values, minimizes van der Waals repulsions, and maximizes noncovalent interactions, with minimal movement away from the averaged atomic coordinates. The result is a single model, the *restrained minimized average structure*. The rms deviation in position from the average of the ensemble is computed for each atom in this model. Deviations are low in regions where models in the ensemble agree well with each other, and deviations are high where the models diverge. Plate 14b shows the final model of thioredoxin (3trx) derived from the full ensemble partially represented in Plate 14a. Colors of the model represent rms deviations of atom positions from the ensemble average, with blue assigned to the smallest deviations, red to the largest, and color across the visible spectrum for intermediate values.

How much structural information must we obtain from NMR in order to derive models like those shown in Plate 14? As summarized in the PDB file header for thioredoxin (3trx), the models shown in Plate 14a were determined from 1983 interproton distance restraints derived from NOE couplings. In addition, the researchers used 52 hydrogen-bonding distance restraints defining 26 hydrogen bonds. Detection of these hydrogen bonds was based on H/D exchange experiments, in which spectra in H_2O and D_2O were compared, and NOEs involving exchangeable H disappeared or were diminished in D_2O. All hydrogen bonds detected in this manner involved exchangeable amide hydrogens. Once these hydrogen bonds were detected in early model construction, they were included in the restraints used in further calculations. Finally, there were 98 Φ and 71 Ψ backbone dihedral-angle restraints, and 72 χ_1 (C_β-C_γ) side-chain dihedral-angle restraints, derived from NOE and J-coupling. Thus the conformation of each of the 33 final thioredoxin models is defined by a total of 2276 restraints. Recall that thioredoxin is a relatively small protein of 105 residues, so these models are based on about 20 restraints per residue.

Finally, for an example of the effect of a single distance restraint on the final model, take another look at Fig. 10.6 and Plate 14c. In Fig. 10.6, one of the NOESY off-diagonal signals is labeled "F89N,δ." This notation means that the correlated nuclei are in phenylalanine-89 (F89), and specifically that the hydrogen on the amide nitrogen is correlated with the hydrogen on one of the C_β carbons, which are on the aromatic ring, *ortho* to the side-chain connection. If you compare this figure with Fig. 10.1, you can identify the diagonal signals: the signal at approximately $\delta = 9.0$ is the amide hydrogen, and the signal at $\delta = 7.3$ is the aromatic hydrogen. Plate 14*c* shows the distance between these two hydrogens in the final averaged model. You can see how this distance restraint confines the conformation of the phenylalanine side chain in this model.

E. PDB files for NMR models

Like crystallographers, NMR spectroscopists share their models with the scientific community by depositing them in the Protein Data Bank. For NMR models, two PDB files—one containing the coordinates of the restrained minimized average structure—usually appear, one containing all models in the ensemble. For thioredoxin, these are 3trx (averaged model) and 4trx (ensemble of 33 models). Many research groups also deposit an accessory file listing all distance restraints used in arriving at the final models.

At first glance, the averaged model would appear to serve most researchers who are looking for a molecular model to help them explain the function of the molecule and rationalize other chemical, spectroscopic, thermodynamic, and kinetic data. On the other hand, you might think that the ensemble and distance-restraint files are of most use to those working to improve structure determination techniques. There are good reasons however, for all researchers to look carefully at the ensemble, as discussed in the next section.

As with PDB files for crystallographic models, NMR coordinate files also include headers containing citations to journal articles about the structure determination work, as well as brief descriptions of specific techniques used in producing the model presented in the file. When you view PDB files in web browsers, the literature citations contain convenient live links to Medline abstracts of the listed articles.

F. Judging model quality

The most obvious criterion of quality of an NMR model is the level of agreement of models in the ensemble. You can make this assessment qualitatively by viewing the superimposed models in a molecular graphics program like the one I will describe the Chapter 11. You should be particularly interested in agreement of the models in regions of functional importance, including catalytically active or ligand-binding sites. The PDB file for the averaged model usually contains information to help you make this assessment. In particular, the data column that contains B-factors for a crystallographic models contain for NMR models the rms deviations from average ensemble coordinate positions. As with B-factors, rms deviations are smaller in the main chain than in side chains. The best quality models currently exhibit main-chain deviations no greater than 0.4 Å, with side-chain values below 1.0 Å.

Many graphics programs will color the averaged model according to these deviations, as illustrated in Plate 14b. This coloring is a useful warning to the user, but it is not as informative as the deviations themselves because the graphics program, depending on its settings, may give colors from blue to red relative to the range of values in the PDB file, regardless of how narrow or wide the range. Thus two models colored by rms deviations may appear similar in quality, but one will have all rms deviations smaller than 3 Å, while the other will have much higher values. You can best compare the quality of two different models by coloring rms deviations on an absolute scale, rather than the relative scale of the contents of a single PDB file.

It is also useful to think about why rms deviations might vary from region to region in the averaged model. In crystallographic models, higher B-factors in sections of a well-refined model can mean that these sections are dynamically disordered in the crystal, and thus moving about faster than the time scale of the data collection. The averaged image obtained by crystallography, just like a photo of a moving object, is blurred. On the other hand, high B-factors may mean static disorder, in which specific side chains or loops take slightly different conformations in different unit cells. Recall that the electron-density map sometimes reveals alternative conformations of surface side chains. Are there analogous reasons for high rms deviations in sections of an NMR model?

The proximate reason for high rms values is the lack of sufficient distance restraints to confine models in the ensemble to a particular conformation. Thus various models converge to different conformations that obey the restraints, but several or many conformations may fill the bill. At the molecular level, there are several reasons why NMR spectroscopy may not provide enough distance restraints in certain regions of the model. The physically trivial reason is spectral resolution. Either resolution may simply not be good enough for all couplings to be resolved and assigned, or some spectral peaks may overlap simply because the structures and environments they represent are practically identical. This problem is worst in the most crowded regions of the spectrum and might persist despite multidimensional simplification. Because a protein NMR spectrum is composed of many similar spin systems, like many valines, some may simply have such similar chemical shifts and couplings that portions of them cannot be distinguished. Alternatively, the width of spectral peaks may be too wide to allow closely spaced peaks to be resolved, and if peak widths exceed coupling constants, correlations cannot be detected. Peak width is inversely related to the rate at which the protein tumbles in solution, and larger proteins tumble more slowly, giving broader peaks. This is unfortunate, of course, because for larger proteins, you need better resolution. But three- and four-dimensional spectra simplify these complex spectra, and the use of one-bond couplings circumvents the correlation problem because one-bond coupling constants are generally much larger than three-bond proton coupling constants. As of this writing, top NMR researchers are claiming that it is now potentially feasible to determine protein structures in the 15- to 35-kDa range (135 to 320 residues) at an accuracy comparable to crystallographic models at 2.5-Å resolution. They see the upper limit of applicability of methods described here as probably 60–70 kDa.

Beyond limitations in resolution are more interesting reasons for high-rms variations within an ensemble of NMR models. Lack of observed distance restraints may in fact mean a lack of structural restraints in the molecule itself. Large variations among the models may be pointing us to the more flexible and mobile parts of the structure in solution. In fact, if we are satisfied that our models are not limited by resolution or peak overlap, we can take seriously the possibility that variations in models indicate real dynamical processes. (Relaxation experiments can be used to determine true flexibility.) Sometimes, mobility may take the form of two or a few different conformations of high-rms regions of the real molecule, each having a long enough lifetime to produce NOE signals. The distance inferred from NOE intensities would be an average figure for the alternative conformations. The ensemble of final models would then reveal conformations that are compatible with averaged distance restraints, and thus point to the longer-lived conformations in solution.

If some or all of the ensemble conformations reveal actual alternative conformations in solution, then these models contain useful information that may be lost in producing the averaged model. If the most important conformations for molecular function are represented in subsets of models within the final ensemble, then an averaged model may mislead us about function. Just like crystallographic models, NMR models do not simply tell us what we would like to know about the inner workings of molecules. Evidence from other areas of research on the molecule are necessary in interpreting what NMR models have to say.

III. Homology models

A. Introduction

I have repeatedly reminded you that protein structure determination is a search for the conformation of a molecule whose chemical composition is known.

Much experience supports the conclusion that proteins with similar amino-acid sequences have similar conformations. These observations suggest that we might be able to use proteins of known structure as a basis for building models of proteins for which we know only the amino-acid sequence. This type of structure determination has been called *knowledge-based modeling* but is now commonly known as *comparative protein modeling* or *homology modeling*. We refer to the protein we are modeling as the *target*, and to the proteins used as frameworks as *templates*.

If, in their core regions, two proteins share 50% sequence identity, the alpha carbons of the core regions can almost always be superimosed with an rms deviation of 1.0 Å or less. This means that the core region of a protein of known structure provides an excellent template for building a model of the core region of a target protein having 50% or higher sequence homology. The largest structural differences between homologous proteins, and thus the regions in which homology models are likely to be in error, lie in surface loops. So comparative modeling is easiest and most reliable in the core regions and the most difficult and unreliable in loops. The structures resulting from homology modeling are, in a sense, low-resolution structures, but they can be of great use, for example, in guiding researchers to residues that might be involved in protein function. Hypotheses about the function of these residues can then be tested by looking at the effects of site-directed mutagenesis. Homology models may also be useful in explaining experimental results, such as spectral properties; in predicting the effects of mutations, such as site-directed mutations aimed at altering the properties or function of a protein; and in designing drugs aimed at disrupting protein function.

Because much of the homology modeling process can be automated, it is possible to develop databases of homology models automatically as genome projects produce new protein sequences. The user of such databases should be aware that automatically generated homology models can often be improved by user intervention in the modeling process (see Section D).

B. Principles

Comparative protein modeling entails these steps (1) constructing an appropriate template for the core regions of the target, (2) aligning the target sequence with the core template and producing a target core model, (3) building surface loops, (4) adding side chains in mutually compatible conformations, and (5) refining the model. In this section, I will give a brief outline of a typical modeling strategy. Although a variety of such strategies have been devised, my discussion is based on the program ProMod and on procedures employed by SWISS-MODEL, a public service homology modeling tool on the World Wide Web. SWISS-MODEL is one of many tools available on the ExPASy Molecular Biology Server operated by the Swiss Institute of Bioinformatics. You will find a link to ExPASy on the CMCC Home Page.

Templates for modeling

The first step in making a comparative protein model is the selection of appropriate templates from among proteins of known structure (determined by crystallography or NMR) that exhibit sufficient homology with the target protein. Even though it might seem that the best template would be the protein having highest sequence homology with the target, usually two or several proteins of high percent homology are chosen. Multiple templates avoid biasing the model toward one protein. In addition, they guide the modeler in deciding where to build loops and which loops to choose. Multiple templates can also aid in the choice of side-chain conformations for the model.

Users find templates by submitting the target sequence for comparison with sequences in databases of known structures. Programs like BLAST or FastA carry out searches for sequences similar to the target. A BLAST score lower than 10^{-5} (meaning a probability lower than one in 100,000 that the sequence similarity is a coincidence), or a FastA score 10 standard deviations above the mean score for random sequences, indicates a potentially suitable template. A safe threshold for automated modeling is 35% homology between target and templates. Next, programs like SIM align the templates. Because many proteins contain more than one chain, and many chains are composed of more than one domain, modelers use databases, extracted from the Protein Data Bank, of single chains. There are also means to find and extract single domains homologous to the target from within larger protein chains.

If only one homolog of known structure can be found, it is used as the template. If several are available, the template will be an averaged structure based on them. The chain most similar to the target is used as the *reference*, and the others are superimposed on it by means of least-squares fitting, with alignment criteria emphasizing those regions that are most similar (referred to as the "conserved" regions because sequence similarities presumably represent evolutionary conservation of sequence within a protein family). To obtain the best alignment, a program like ProMod starts by aligning alpha carbons from only those regions that share the highest homology with the reference. Alpha carbons from the other (nonreference) templates are added to the aligned, combined template if they lie within a specified distance, say 3.0 Å, of their homologous atoms in the reference. The result is called a structurally corrected multiple-sequence alignment.

Modeling the target core

To build the core regions of the target protein, its sequence is first aligned
with that of the template or, if several templates are used, with the structurally
corrected multiple sequence alignment. The procedure aligns the target with
all regions, including core and loops, that give high similarity scores, with the
result that the target aligns with the core regions of all models, whereas loop
regions of the target align only with individual models having very similar
loop sequences. This means that subsequent modeling processes will take advantage not only of the general agreement between target and all templates in
core regions but also of the specific agreement between target and perhaps a
single template that shares a very similar loop sequence.

With this sequence alignment, a backbone model of the target sequence is
threaded or folded onto the aligned template atoms, producing a model of the
target in the core regions and in any loops for which a highly homologous
template loop was found. When multiple templates are used, the target atom
positions are the best fit to the template atom positions for a target model that
keeps correct bond lengths and angles, perhaps weighted more heavily toward
the templates of highest local sequence homology with the target. As for
residue side chains in the averaged template, in SWISS-MODEL, side-chain
atom positions are averaged among different templates and added so that space
is occupied more or less as in the templates. After averaging, the program selects, for each side chain, the allowed side-chain conformation or *rotamer* that
best matches the averaged atom positions.

Modeling loops

At this stage, we have a model of the core backbone in which the atom positions are the average of the atom positions of the templates, and in which some
core side chains are included. Perhaps some loops are also modeled, if one or
more templates were highly homologous to the target. But most of the surface
loops are not yet modeled, and in most cases, the templates give us no evidence about their structures. This is because the most common variations
among homologous proteins lie in their surface loops. These variations include differences in both sequence and loop size. The next task is to build reasonable loops containing the residues specified by the target sequence.

One approach is to search among high-resolution (2.5-Å) structures in the
Protein Data Bank for backbone loops of the appropriate size (number of
residues) and end-point geometry. In particular, we are looking for loops whose
conformation allows their own neighboring residues to superimpose nicely on
the target loop's neighboring residues in the already modeled target core region. To hasten this search, modelers extract and keep loop databases from the

PDB, containing the alpha-carbon coordinates of all loops plus those of the four neighboring residues, called "stem" residues, on both ends of each loop.

To build a loop, a modeling program selects loops of desired size and scores them according to how well their stem residues can be superimposed on the stem residues of the target core, aligning each prospective loop as a rigid body. The program might search for loops whose rms stem deviations are less than 0.2 Å and, finding none, might search again with a criterion of 0.4 Å, continuing until a small number, say five, candidates are found. The target loop is then modeled, its coordinates based on the average of five database loops with lowest rms-deviation scores. This process builds only an alpha-carbon model, so the amide groups must be added. This might involve another search through even higher-resolution structures in the PDB, looking for short peptides (not necessarily loops) whose alpha carbons align well with short stretches of the current loop atoms. Again, if several good fits are found, the added atoms are modeled on their average coordinates, thus completing the backbone of our model.

Modeling side chains

Our model now consists of a complete backbone, with side-chains only present on those residues where templates and target are identical. In regions of high sequence homology, target side chains nonidentical to templates might be modeled on the template side chains out to the gamma atoms. The remaining atoms and the remaining full side chains called for by the target sequence are then added, using rotamers of the side chains that do not clash with those already modeled, or with each other.

Refining the model

Our model now contains all the atoms of the known amino-acid sequence. Because many atom positions are averages of template positions, it is inevitable that the model harbors clashing atoms and less-than-optimal conformations. The end of all homology modeling is some type of structure refinement, including energy minimization. SWISS-MODEL uses the program GROMOS to idealize bond lengths and angles, remove unfavorable atom contacts, and allow the model to settle into lower-energy conformations lying near the final modeled geometry. The number of cycles of energy minimization is limited so that the model does not drift too far away from the modeled geometry. In particular, loops tend to flatten against the molecular surface upon extended energy minimization, probably because the model's energy is being minimized in the absence of simulated interactions with surrounding water.

A modeling process like the one I have described is applied automatically—but with the possibility of some user modifications—to target sequences submitted to the SWISS-MODEL server at ExPASy. This tool also allows for intervention at various stages, during which the user can apply special knowledge about the target protein or can examine and adjust sequence or structural alignments. Swiss-PdbViewer—an excellent program for viewing, analyzing, and comparing models—is specifically designed to carry out or facilitate these interventions for SWISS-MODEL users, thus allowing a wider choice of tools for template searching, sequence alignment, loop building, threading, and refinement. You will learn more about Swiss-PdbViewer in Chapter 11.

C. Databases of homology models

As a result of the many genome-sequencing projects now under way, an enormous number of new structural genes (genes that code for proteins) are being discovered. With the sequences of structural genes comes the sequences of their product proteins. Thus new proteins are being discovered far faster than crystallographers and NMR spectroscopists can determine their structures. Although homology models are likely to be less accurate than those derived from experimental data, they can be obtained rapidly and automatically. Though they cannot guide detailed analysis of protein function, these models can guide further experimental work on a protein, such as site-directed mutagenesis to pinpoint residues essential to function. The desirability of at least "low-resolution" structures of new proteins has led to initiatives to automatically model all new proteins as they appear in specified databases. Thus the number of homology models has already exceeded the number of experimentally determined structures in the Protein Data Bank.

One of the first automated modeling efforts was carried out in May of 1998. Called 3D-Crunch, the project entailed submission of all sequences in two major sequence databases, SWISS-PROT and trEMBL, to the SWISS-MODEL server. The result was about 64,000 homology models (compared to about 9,000 models in the Protein Data Bank at the time), which are now available to the public through the SWISS-MODEL Repository. A versatile searching tool allows users to find and download final models. Alternatively, users can download entire modeling projects, containing the final coordinates of the homology model along with aligned coordinates of the templates. These project files enable the user to carry the modeling project beyond the automated process and to use other tools or special knowledge about the protein to further improve the model. Project files are most conveniently opened in Swiss-PdbViewer, which provides both built-in modeling tools and interfaces to additional programs for further improvement of models. Plate 20 shows a modeling project returned from SWISS-MODEL. For more discussion of this plate, see the next section and the plate legend.

D. Judging model quality

Note that the entire comparative protein-modeling process is based on structures of proteins sharing sequence homology with the target, and that no experimental data about the target is included. This means that we have no criteria, such as R-factors, that allow us to evaluate how well our model explains experimental observations about the molecule of interest. There are no experimental observations. At the worst, in inept hands, a homology modeling program is capable of producing a model of any target from even the most inappropriate template. Even the most respected procedures may produce dubious models. By what criteria do we judge the quality of homology models?

We would like to ask whether the model is *correct*. We could say it is correct if it agrees with the actual molecular structure. But we do not have this kind of assurance about any model, even one derived from experiment. It is more reasonable for us to define *correct* as agreeing to within experimental error with an experimental (X-ray or NMR) structure. Most of the time, we accept a homology model because we do not have an experimental structure, but not always. Researchers working to improve modeling methods often try to model known structures starting from related known structures. Then they compare the model with the known structure to see how the modeling turned out. In this situation, a correct model is one in which the atom positions deviate from those of the experimental model by less than the uncertainty in the experimental coordinates, as assessed by, for example, a Luzzati plot (Chapter 8, Section I.B). Areas in which the model is incorrect are arenas for improving modeling tools.

Typically, however, we want to use a homology model because it is all we have. In some cases, even without an experimental model for comparison, we can recognize incorrectness. A model is incorrect if it is in any sense impossible. What are signs of an incorrect model? One is the presence of hydrophobic side chains on the surface of the model, or buried polar or ionic groups that do not have their hydrogen-bonding or ionic-bonding capabilities satisfied by neighboring groups. Another is poor agreement with expected values of bond lengths and angles. Another is the presence of unfavorable noncovalent contacts or "clashes." Still another is unreasonable conformational angles, as exhibited in a Ramachandran plot (Chapter 8, Section I.A). We know that high-quality models from crystallography and NMR do not harbor these deficiencies, and we should not accept them in a homology model. Many molecular graphics programs can compute deviations from expected bond geometry;

highlight clashes with colors, dotted lines, or overlapping spheres; and display Ramachandran diagrams, thus giving us immediate visual evidence of problems with models. We can also say that the model is incorrect if the sequence alignment is incorrect or not optimal. The details of sequence comparison are beyond the scope of this book, but we can test the alignment of target with templates by using different alignment procedures, or by altering the alignment parameters to see if the current alignment is highly sensitive to slight changes in method. If so, it should shake our confidence in the model.

Beyond criteria for correctness, we can also ask how well the model fits its templates. The rms deviation of model from template should be very small in the core region. If not, we say that the model is *inaccurate*. An inaccurate model implies that the modeling process did not go well. Perhaps the modeling program simply could not come up with a model that aligns well with the coordinates of the template or templates. Perhaps during energy minimization, coordinates of the model drifted away from the template coordinates. Another possibility is poor choice of templates. For instance, occasionally a crystallographic model is distorted by crystal contacts, or an NMR template model is distorted by the binding of a salt ion. If we unwittingly use such models as templates, energy refinement in the absence of the distorting effect would introduce inaccuracy, as defined here, while perhaps actually improving the model. A good rule of thumb is that if the templates share 30–50% homology with the target, rms differences between final positions of alpha carbons in the model and those of corresponding atoms in the templates should be less than 1.5 Å. But it is also essential to look at the template structures and make sure that they are really appropriate. An NMR structure of an enzyme-cofactor complex is likely to be a poor model for a homologous enzyme in the absence of the cofactor.

The rms deviations only apply to corresponding atoms, which means mainly the core regions. Loop regions often cannot be included in such assessment because there is nothing to compare them to. Again, we should demand correctness, that is, the lack of unfavorable contacts or conformations. But beyond this kind of correctness, our criteria are limited. If surface loops contain residues known to be important to function, we must proceed with great caution in using homology models to explain function.

If the model appears to be correct (not harboring impossible regions like clashes) and accurate (fitting its templates well), we can also ask if it is *reasonable*, or in keeping with expectations for similar proteins. Researchers have developed several assessments of reasonableness that can sometimes signal problems with a model or specific regions of a model. One is to sum up the probabilities that each residue should occur in the environment in which it is found in the model. For all Protein Data Bank models, each of the 20 amino acids has a certain probability of belonging to one of the following classes:

solvent-accessible surface, buried polar, exposed nonpolar, helix, sheet, or turn. Regions of a model that do not fit expectations based on these probabilities are suspect.

Another criterion of reasonableness is to look at how often pairs of residues interact with each other in the model in comparison to the same pairwise interactions in templates or proteins in general. The sum of pairwise potentials for the model, usually expressed as an "energy" (smaller is better) should be similar to that for the templates. One form of this criterion is called *threading energy*. Threading energy indicates whether the environment of each residue matches what is found for the same residue type in a representative set of protein folds. Such criteria ask, in a sense, whether a particular stretch of residues is "happy" in its three-dimensional setting. If a fragment is "unhappy" by these criteria, then that part of the model may be in error.

To be meaningful, all of these assessments of reasonableness of a model must be compared with the same properties of the templates. After all, the templates themselves, even if they are high-quality experimental structures, may be unusual in comparison to the average protein.

Homology model coordinate files returned from SWISS-MODEL contain, in the *B*-factor column, a *confidence factor*, which is based on the amount of structural information that supports each part of the model. Actually, it would be better to call this figure an *uncertainty factor*, or a *model B-factor*, because a high value implies high uncertainty, or low confidence, about a specific part of the model. (Recall that higher values of the crystallographic *B*-factor imply greater uncertainty in atom positions.) The model *B*-factor for a residue is higher if fewer template structures were used for that residue. It will also be higher for a residue whose alpha-carbon position deviates by more than a specified distance from the alpha carbon of the corresponding template residue. This distance is called the *distance trap*. In SWISS-MODEL (or more accurately, in ProModII, the program that carries out the threading at SWISS-MODEL), the default distance trap is 2.5 Å, but the user can increase or decrease it. However, if the user increases the distance trap, all of the model *B*-factors increase, so they still reflect uncertainty in the model, even if the user is willing to accept greater uncertainty. Finally, all atoms that are built without a template, including loops for which none of the template models had a similar loop size or sequence, are assigned large model *B*-factors, reflecting the lack of template support for those parts of the model. Computer displays of homology models can be colored by these model *B*-factors to give a direct display of the relative amount of information from X-ray or NMR structures that was used in building the model (Chapter 11, Section III.E). Plate 20 shows a homology model and its templates. The target model is colored by the model *B*-factors assigned by SWISS-MODEL. The templates are black and gray. With this color scheme, it is easy to distinguish the parts of the model in which we can have the most confidence. Blue regions were built on more templates and fit the templates better. Red regions were built completely from loop databases, without template contributions. Colors of the visible spectrum between blue and red may align well with none or only a subset of the templates. (For more information about these proteins, see the legend to Plate 20.)

IV. Other theoretical models

Structural biologists produce other types of theoretical models as they pursue research in various fields. Work that produces new models includes developing schemes to predict protein conformation from sequence; attempts to simulate folding or other dynamic processes; and attempts to understand ligand binding by building ligands into binding sites and then minimizing the energy of the resulting combined model. As of December 1998, the Protein Data Bank listed about 200 theoretical models among its approximately 9000 models. Many of these were homology models. According to on-line documentation, "The Protein Data Bank (PDB) is an archive of *experimentally determined* three-dimensional structures. . . ." The presence of theoretical models in the Protein Data Bank was only a temporary measure due to the lack of a data bank for homology and other theoretical models. An international database for theoretical models is now being established, and should be open to the public sometime in 1999. It will probably include models from the SWISS-MODEL Repository, theoretical models now in the PDB, and all other types of theoretical models. Look for links to model databases on the CMCC Home Page.

My goals in this chapter were to make you aware of the variety of model types available to the structural biologist, to give you a start toward understanding other methods of structure determination, and to guide you in judging the quality of noncrystallographic models, primarily by drawing your attention to analogies between criteria of quality in crystallography.

Models are are not molecules observed. No matter how they are obtained, before we ask what they tell us, we must ask how well macromolecular models fit with other things we already know. A model is like any scientific theory: it is useful only to the extent that it supports predictions that we can test by experiment. Our initial confidence in it is justified only to the extent that it fits what we already know. Our confidence can grow only if its predictions are verified.

11 Tools for Studying Macromolecules

I. Introduction

There is an old line about a dog who is finally cured of chasing cars—when it catches one. What to do now? In this chapter, I discuss what to do when you catch a protein. My main goal is to inform you about some of the tools available for studying protein models and to suggest strategies for learning your way around the unfamiliar terrain of a new protein. I will begin with a very brief glimpse of the computations that underlie molecular graphics displays. Then I will take you on a tour of molecular modeling by detailing the features present on most modeling programs. Finally, I will briefly introduce other computational tools for studying and comparing proteins. My emphasis is on tools that you can use on your personal computer.

II. Computer models of molecules

A. Two-dimensional images from coordinates

Computer programs for molecular modeling provide an interactive, visual environment for displaying and exploring models. The fundamental operation of computer programs for studying molecules is producing vivid and understandable displays—convincing images of molecules. Although the details of programming for graphics displays vary from one program (or programming language, or computer operating system) to another, they all produce an image according to the same geometric principles.

A display program uses a file of atomic coordinates to produce a drawing on the screen. Recall that a PDB coordinate file contains a list of all atoms located by crystallographic, NMR, or theoretical analysis, with coordinates x, y, and z for each atom. When the model is first displayed, the coordinate system is usually shifted by the modeling program so that the origin is the center of the model. This origin lies at the center of the screen, becoming the origin of a new coordinate system, the *screen coordinates*, which I will designate x_s, y_s, and z_s. The x_s-axis is displayed horizontally, y_s is vertical, and z_s is perpendicular to the computer screen. As the model is moved and rotated, the screen coordinates are continually updated.

The simplest molecular displays are stick models with lines connecting atoms, and atoms simply represented by vertices where lines meet (see Plate 15a for examples of various model displays or *renderings*). It is easy to imagine a program that simply plots a point at each position (x_s, y_s, z_s) and connects the points with lines according to a set of instructions about connectivity of atoms in amino acids.

But the computer screen is two-dimensional. How does the computer plot in three dimensions? It doesn't; it plots a projection of the three-dimensional stick model. Mathematically, projecting the object into two dimensions involves some simple trigonometry, but graphically, projecting is even simpler. The program plots points on the screen at positions $(x_s, y_s, 0)$; in other words, the program does not employ the z_s coordinate in producing the display. This produces a projection of the molecule on the $x_s y_s$ plane of the screen coordinate system, which is the computer screen itself (see Fig. 11.1). You can imagine this projection process as analogous to casting a shadow of the molecule on the screen by holding it behind the screen and lighting it from behind. Another analogy is the projected image of a tree onto the ground made by its leaves when they fall during a cold windless period. The program may use the z-coordinate to produce shading or perspective, or to allow foreground object to cover background objects, and thus to make the display look three-dimensional.

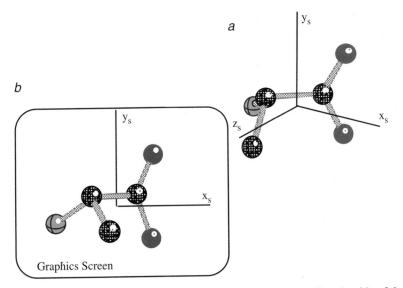

a

b

Graphics Screen

Figure 11.1 Geometry of projection. (*a*) Model viewed from off to the side of the screen coordinate system. Each atom is located by screen coordinates x_s, y_s, and z_s. (*b*) Model projected onto graphics screen. Each atom is displayed at position $(x_s, y_s, 0)$, producing a projection of the model onto the $x_s y_s$-plane, which is the plane of the graphics screen.

B. Into three dimensions: Basic modeling operations

The complexity of a protein model makes it essential to display it as a three-dimensional object and move it around (or move our viewpoint around within it). The first step in seeing the model in three dimensions is rotating it, which gives many three-dimensional cues and greatly improves our perception of it. Rotating the model to a new orientation entails calculating new coordinates for all the atoms and redisplaying by plotting on the screen according to the new (x_s, y_s) coordinates.

The arithmetic of rotation is fairly simple. Consider rotating the model by θ degrees around the x_s (horizontal) axis. It can be shown that this rotation transforms the coordinates of point p, $[x_s(p), y_s(p), z_s(p)]$, to new coordinates $[x'_s(p), y'_s(p), z'_s(p)]$ according to these equations:

$$
\begin{aligned}
x'_s(p) &= x_s(p) \\
y'_s(p) &= [y_s(p) \cdot \cos\theta] - [z_s(p) \cdot \sin\theta] \\
z'_s(p) &= [y_s(p) \cdot \sin\theta] + [z_s(p) \cdot \cos\theta].
\end{aligned}
\qquad (11.1)
$$

Notice that rotating the model about the x_s-axis does not alter the x_s coordinates, $(x'_s(p) = x_s(p))$, but does change y_s and z_s. A similar set of equations provides for rotation about the y_s- or z_s-axis. If we instruct the computer to rotate the model around the x_s-axis by θ degrees, it responds by converting all coordinates (x_s, y_s, z_s) to new coordinates (x'_s, y'_s, z'_s) and plotting all points on the screen at the new positions $(x'_s, y'_s, 0)$. Graphics programs allow so-called real-time rotation in which the model appears to rotate continuously. This requires fast computation because to produce what looks like smooth rotation, the computer must produce a new image about every 0.05 second. So in literally the blink of an eye, the computer increments θ by a small amount, recalculates coordinates of all displayed atoms, using equations such as (11.1), and redraws the screen image. Fast repetition of this process gives the appearance of continuous rotation.

C. Three-dimensional display and perception

Complicated models become even more comprehensible if seen as three-dimensional (3-D) objects even when not in motion. So graphics display programs provide some kind of full-time 3-D display. This entails producing two images like the stereo pairs used in this book, and presenting one image to the left eye and the other to the right eye, which is the function of a stereo viewer. The right-hand view is just like the left-hand view, except that it is rotated about 5° about its y_s axis (counterclockwise, as viewed from above). Molecular modeling programs display the two images of a stereo pair side by side on the screen, for viewing in the same manner as printed pairs. (For tips on viewing in stereo, see the information above Plate 1 or the CMCC Home Page.)

With proper hardware, modeling programs can produce full-screen 3-D objects. The technique entails flashing full-size left and right views alternately at high speed. The viewer wears special glasses with liquid-crystal lenses that alternate rapidly between opaque and transparent. When the left-eye view is on the screen, the left lens is clear and the right lens is opaque. When the screen switches to the right-eye view, the right lens becomes clear and the left becomes opaque. In this manner, the left eye sees only the left view, and the right eye sees only the right view. The alternation is fast, so switching is undetectable, and a full-screen 3-D model appears to hang suspended before the viewer.

Both types of stereo presentation mimic the appearance of objects to our two eyes, which produce images on the retina of objects seen from two slightly different viewpoints. The two images are rotated about a vertical axis located at the current focal point of the eyes. From the difference between the two images, called *binocular disparity*, we obtain information about the relative depth of objects in our field of view. For viewers with normal vision, two pictures with a 5° binocular disparity, each presented to the proper eye, gives a convincing three-dimensional image.

Unfortunately, a small but significant percentage of people cannot obtain depth information from binocular disparity alone. In any group of 20 students, there is a good chance that one or more will not be able to see a three-dimensional image in printed stereo pairs. In the real world, we decode depth not only by binocular disparity but also by relative motion, size differences produced by perspective, overlap of foreground objects over background objects, effects of lighting, and other means. Molecular modeling programs provide depth information to a wider range of viewers by providing depth cues in the form of shading, perspective, and movement, as described in the next section.

D. Types of graphical models

Modeling and graphics programs provide many ways to *render* or represent a molecular model. The simplest is the wireframe model. A wireframe model of three strands of beta sheet from cytochrome $b5$ (PDB 3b5c[1]) is shown in Plate 15a. Although other renderings have greater visual impact, the wireframe model is without question the most useful when you are exploring a model in detail, for several reasons. First, you can see the model and see through it at the same time. Foreground and background parts of the model do not block your view of the part of the model on which you center your attention. Second, in wireframe models, you can see atom positions, bond angles, and torsional angles clearly. Third, wireframe models are the simplest and hence the fastest for your computer to draw. This means that, as you rotate or zoom a model, the atom positions are recalculated and the images redrawn faster, so that the model moves more rapidly and smoothly on the screen. Rapid and smooth motion give the model a tangible quality that improves your ability to grasp and understand it. Combined with colorings that represent element types, or other properties like B-factors or solvent accessibility (Section III.H), wireframe models can be very informative. There is one disadvantage: wireframe models do not contain obvious depth cues; therefore, they are best viewed in stereo.

Other renderings, some of which are illustrated in Plate 15, have their strengths and weaknesses. All of them make vivid illustrations in textbooks and on web pages, when conveying fine structural details is not necessary. It takes programs longer to draw these more complex models, so they might not

[1]F. S. Mathews, M. Levine, and P. Argos, *Nature,* **233,** 15, 1971. Atomic coordinates (PDB 3b5c, revised 1990) obtained from the Protein Data Bank (Chapter 7).

move as smoothly on the screen. Ball-and-stick models are similar to wireframe models, but they provide better depth cues, because they are shaded and appear as solid objects that obscure objects behind them. Plate 15*b* shows a ball-and-stick model of the beta strands shown in (*a*).

The same atoms are shown again in Plate 15*c* as a space-filling model, which gives a more realistic impression of the density of a model, and of the extent of its surface. The surface of a space-filling model can be either the van der Waals (shown) or the solvent-accessible surface, and can be colored according to various properties, like surface charge. Space-filling models are excellent for examining the details of atomic contacts within and between models. Space-filling models of an enzyme and its inhibitor appear to fit each other like hand and glove. It is impossible to see exact atom positions and angles in space-filling models.

Various cartoon renderings, like the ribbon diagram in Plate 15*d*, show main chains as parallel strands or solid ribbons, perhaps with arrows on strands of pleated sheet to show their direction, and barrel-like alpha helices. These are some of the most vivid renderings and are excellent for giving an overview of protein secondary and tertiary structure. Combined images, for example, with most of the molecule in cartoon form and an important binding site in ball and stick, make nice printed illustrations that guide the reader from the forest of the whole molecule to the trees of functionally important structural details. On the other hand, rotation of these images on a personal computer can be quite choppy and slow.

III. Touring a typical molecular modeling program

As you might infer from a brief description of the computing that underlies molecular graphics, the computer must be fast. But today's personal computers have gradually come up to the task of dealing with quite large and complex graphics. The current generation of personal computers, using fast conventional processors, so-called RISC (reduced-instruction-set computer) processors, or two or more conventional processors in parallel, can allow you to conduct highly satisfying and informative expeditions into the hearts of the most complex macromolecules.

The basic operations of projecting and rotating a screen image of the molecular model underlie all molecular graphics programs. Upon these operations are built many tools for manipulating the display. These tools give viewers the feeling of actively exploring a concrete model. Now I will discuss tools commonly found in modeling programs.

Although this is a general discussion of modeling tools, I will frequently mention as an example an excellent molecular viewing and modeling program that you can obtain free of charge in IBM-PC clone, Linux, Macintosh, or Silicon Graphics versions. My example is Swiss-PdbViewer (SPV), an easy-to-learn yet very powerful molecular viewing and modeling tool. Plate 16 provides a picture of a personal computer screen during use of SPV, with a description of its main sets of functions. If you currently own no modeling program and want to learn to use a program that will allow you start easily, yet grow painlessly into very sophisticated modeling, SPV is an excellent choice. It far exceeds another popular free program, RasMol, in ease of use, speed, and analytical power. In fact, it compares favorably with costly commercial programs. To learn how to download this program from the World Wide Web and to find a complete, self-guiding tutorial for modeling beginners, see the CMCC Home Page. Whatever the program you decide you use, search its documentation or the World Wide Web for tutorials on how to get started with it. There is no faster way to learn molecular exploration and analysis than by a hands-on tutorial. The CMCC Home Page also contains links to tutorials for other widely used molecular graphics programs.

A. Importing and exporting coordinate files

As indicated earlier, image display and rotation requires rapid computing. Reading coordinates from a text file like a PDB file is slow because every letter or number in the file must be translated into binary or machine language for the computer's internal processing. For this reason, graphics programs work with a machine-language version of the coordinate file, which they can read and recompute faster than a text file. So the first step in exploring a model is usually converting the PDB file to binary form. With almost all modeling programs, this operation is invisible to the user.

Often users produce revealing views of the model and wish to use the coordinates in other programs, such as energy calculations or printing for publication. For this purpose, molecular modeling programs include routines for writing coordinate files in standard formats like PDB, using the current binary model coordinates as input.

B. Loading and saving models

The coordinate files produced by graphics programs can be loaded and saved just like any other files. As the model is manipulated or altered, coordinates are updated. At any point, the user can save the current model, or replace it with one saved earlier. To facilitate studying intermolecular interactions, modeling programs can handle several models at once, perhaps more than ten at one time. The models currently in memory can be viewed and manipulated individually or together. You can save them with their their current relative orientations so that you can resume a complicated project later. Although almost all macromolecular model coordinates are available in PDB format, modeling programs provide for loading and saving files in a variety of formats. These might include Cambridge Structure Database format, used by the largest database for small molecules, or others from among the dizzying list of small-molecule formats used by organic and inorganic chemists. Other special file formats include those used by energy minimization, dynamics simulation, and refinement programs, most of which run only on computers that are faster and more powerful than personal computers. Full-featured modeling programs like SPV usually provide for exporting coordinates for these powerful programs, shipping them over networks for completion of complex computing tasks, retrieving the resulting model files, and converting them back to PDB format for viewing. As personal computers get faster, these transfers will become unnecessary.

C. Viewing models

Standard viewing commands allow users to rotate the model around screen x_s, y_s, and z_s axes and move (translate) the model for centering on areas of interest. Viewers magnify or zoom the model by moving it along the z_s-axis toward the viewer, or by using a command that magnifies the image without changing coordinates by simply narrowing the viewing angle. "Clip" or "slab" commands simplify the display by eliminating foreground and background, producing a thin slab of displayed atoms. With most programs, including SPV, all of these operations are driven by a mouse or other point-and-drag device (trackball, for example), after selecting the desired operation by buttons (top of graphics window in Plate 16) or key commands.

Plate 17 shows two views of the small protein cytochrome $b5$ (PDB 3b5c) an electron-transport protein containing an iron-heme prosthetic group (shown in green with red iron at the center). In (a), you are looking into the heme binding pocket, with much of the protein in the background. Even in stereo, the background clutter makes it difficult to see the heme environment clearly. In (b), the background is removed using a "slab" command to give a clearer view of the heme and protein groups above and below it.

The program clips by displaying only those atoms whose current z_s-coordinates lie within a specified range, which is chosen visually by sliding front and rear clipping planes together until unwanted background and foreground atoms disappear.

Centering commands allow the user to select an atom to be made the center of the display. Upon selection by pointing to the atom and clicking a mouse, or by naming the atom, the program moves the model so as to center the desired atom on the screen and within the viewing slab, and also to make it the center of subsequent rotations. For example, SPV provides a very handy centering feature. Simply pressing the *help* key centers the part of the model currently on display, and resizes it to fit the screen. It is not unusual for novice viewers to accidentally move the model completely out of view and be unable to find it. Nothing is more disconcerting to a beginner than completely losing sight of the model. When the model disappears, it may be off to the side of the display, above or below the display, or still centered, but outside the slab defined by clipping planes. In any event, automatic recentering can often help viewers find the model and regain their bearings. As a last resort, there is usually a reset command, which brings model and clipping planes back to starting positions. The viewer pays a price for resetting, losing the sometimes considerable work of finding a particularly clear orientation for the model, centering on an area of interest, and clipping away obscuring parts of the structure.

Viewing commands usually also include selection of stereo or mono viewing and offer various forms of *depth cueing* to improve depth perception, either by mimicking the effects of perspective (front of model larger than rear), shading (front of model brighter than rear), or rocking the model back and forth by a few degrees of rotation about the y_s-axis.

D. Editing and labeling the display

A display of every atom in a protein is often forbidding and incomprehensible. Viewers are interested in some particular aspect of the structure, such as the active site or the path of the backbone chain, and may want to delete irrelevant parts of the model from the display. Display commands allow viewers to turn atoms on and off. Atoms not on display continue to be affected by rotation and translation, so they are in their proper places when redisplayed. Viewers might eliminate specific atoms by pointing to them and clicking a mouse, or they might eliminate whole blocks of sequence by entering residue numbers. They may display only alpha carbons to show the folding of the protein backbone (refer to Plate 5), or only the backbone and certain side chains to pinpoint specific types of interactions.

Editing requires knowledge of how atoms are named in the coordinate file, which is often, but not always, the same as PDB atom labels (Chapter 7, Section VIII). Thus viewers can produce an alpha-carbon-only model by limiting

the display to atoms labeled CA. Each program has its own language for naming atoms, residues (by number or residue name), distinct chains in the model
(like the α and β chains of hemoglobin), and distinct models. Viewers must
master this language in order to edit displays efficiently. In SPV, editing the
view is greating simplified by a Control Panel (Plate 16, right side) that provides a scrollable list of all residues in the PDB file. A menu at the top allows
you to switch to other PDB files if you are currently viewing more than one
model. You can select, display, label, and color residues, and add surface displays as well, all with mouse operations.

Most programs provide powerful selection tools to allow you to pick specific parts of the model for displaying, labeling, or coloring. For example,
SPV provides menu commands for selecting residues by type (for example,
select all histidines, or select water, or select heteromeric group), property (select acidic residues), or secondary structure (select helices or beta strands);
surface or buried residues; residues with *cis*-peptide bonds; problem residues
in a model, like those with Φ and Ψ values outside the range of allowed values (Chapter 8, Section II.A), or residues making clashes; neighbors of the
current selection (select within 4.5 Å of current selection); and residues in
contact with another chain, to name just a few. I made the view in Plate 16 in
just a few seconds by selecting the heme group and then displaying neighbors
within 5.0 Å of the current selection.

Even with an edited model, it is still easy for viewers to lose their bearings.
Label commands attach labels to specified atoms, signifying element, residue
number, or name. Labels like the one for PHE-58 in Plate 16 float with the atom
during subsequent viewing, making it easy to find landmarks in the model.

E. Coloring

Although you may think that color is merely an attractive luxury, adding color
to the display makes it dramatically more understandable. Most programs
allow atoms to be colored manually, by selecting a part of the model and then
choosing a color from a color wheel or palette. Additional color commands
allow the use of color to identify elements or specific residues, emphasize
structural elements, or display properties. For example, SPV provides, among
others, commands for coloring the currently selected residues by CPK color
(as in Plate 16), residue type, *B*-factor, secondary structure (different colors
for helix, sheet, and turns), chain (a different color for each monomer in an
oligomeric protein), layer (different color for each model currently on
display), solvent accessibility, threading or force-field energies (Chapter 10,
Section III.D), and various kinds of model problems.

Warning: Coloring by *B*-factor uses whatever information is in the *B*-factor column of the PDB file. To interpret the colors that result, it is crucial for you to know what kind of model you are viewing, because this information tells you different things about different kinds of models. Coloring a good crystallographic model by *B*-factor reveals relative uncertainty in atom positions due to static or dynamic disorder (Chapter 8, Section II.C). Coloring an averaged NMR model by "*B*-factor" actually reveals relative rms deviation of atom positions from the average positions of the corresponding atoms in the ensemble of NMR models (Chapter 10, Sections II.E and II.F). And coloring a homology model "by *B*-factor" distinguishes parts of the model according to how much support exists for the model in the form of X-ray or NMR structural information (Chapter 10, Section III.D).

Combined with selecting tools, color commands can be powerful tools simply for finding features of interest. If you are interested in cysteines in a protein, you can select all cysteines, choose a vivid color for them, and immediately be able to find them in the forest of a large protein.

F. Measuring

Measurements are necessary in identifying interactions within and between molecules. In fact, noncovalent interactions like hydrogen bonds are defined by the presence of certain atoms at specified distances and bond angles from each other. In Plate 16, yellow dotted lines connect two oxygen atoms (red) of a heme carboxyl group to the amide nitrogen and the side chain oxygen oxygen of a serine residue. The distances between atoms are displayed and are approximately the distance expected for oxygen atoms involved in O-H—O or N-H—O hydrogen bonds. (Recall that crystallographic studies of proteins do not usually reveal hydrogen atoms.)

Modeling programs allow display of distances, bond angles, and dihedral angles between bonded and nonbonded atoms. These measurements float on dotted lines within the model (just like labels) and are often active; that is, they are continually updated as the model is changed, as described in the next section.

G. Exploring structural change

Modeling programs allow the viewer to explore the effects of various changes in the model, including conformational rotation, change in bond length or angle, and movement of fragments or separate chains. Used along with active measurements, these tools allow viewers to see whether side chains can move to new positions without colliding with other atoms, or to examine the range of possible movements of a side chain. In Plate 16, a change of torsion angles is in progress for phenylalanine-58. The side chain is yellow in the left view because the mouse pointer is near it in the Align window. In its current conformation, the side chain is in collision with the heme, as shown by pink "clash" lines.

Like rotation and translation, changes of model conformation, bond angles, or bond lengths are reflected by changes in the coordinate file. The changes are tentative at first, while users explore various alterations of the model. After making changes, users have a choice of saving the changes, removing the changes, or resetting in order to explore again from the original starting point. In Plate 16, notice that the torsions button is darkened, showing that the operation is in progress, and that the user will have an opportunity to keep or discard the changes being made.

H. Exploring the molecular surface

Stick models of the type shown in Plate 15a are the simplest and fastest type of model to compute and display because they represent the molecule with the smallest possible number of lines drawn on the screen. Stick models are relatively open, so the viewer can see through the outer regions of a complex molecule into the interior or into the interface between models of interacting molecules. But when the viewer wants to explore atomic contacts, a model of the molecular surface is indispensable.

Published structure papers often contain impressive space-filling computer images of molecules, with simulated lighting and realistic shadows and reflections. These images require the computer to draw hundreds of thousands of multicolored lines, and so the computer cannot redraw such images fast enough for continuous movements. Some of these views require hours to draw just once. Although such views show contacts between atomic surfaces, they are not practical for exploring the model interactively. They are used primarily as snapshots of particularly revealing views.

How then can you study the surface interactively? The most common compromise is called a dotted surface (Plate 18), in which the program displays dots evenly spaced over the surface of the molecule. This image reveals the surface without obscuring the atoms within and can be redrawn rapidly as the viewer manipulates the model. Several types of surfaces can be computed, each with its own potential uses. One type is the van der Waals surface, in which all dots lie at the van der Waals radius from the nearest atom, the same as the surface of space-filling models. This represents the surface of contact between nonbonded atoms. Any model manipulations in which van der Waals surfaces penetrate each other are sterically forbidden. Van der Waals surfaces show packing of structural elements with each other, but the display is complicated because all internal and external atomic surfaces are shown.

Another useful surface display is the *solvent-accessible surface*, which shows all parts of the molecule that can be reached by solvent (usually water) molecules. This display omits all internal atomic surfaces, including crevices that are open to the outside of the model, but too small for solvent to enter. Some modeling programs, like SPV, contain routines for calculating this surface, whereas others can take as input the results of surface calculations from widely available programs. Calculating the solvent-accessible surface entails simulating the movement of a sphere, called a "probe," having the diameter of a solvent molecule over the entire model surface, and computing positions of evenly spaced dots wherever model and solvent come into contact. Plate 18 shows the van der Waals surface and the accessible surface for the heme group in cytochrome $b5$. The van der Waals surface is the inner surface that completely surrounds the heme. The outer dots around the heme carboxyls are the accessible surface. It is clear from this view that most of the heme is buried within the protein and not accessible to the solvent.

Carrying out the same simulation that produces solvent-accessible surface displays, but locating the dots at the center of the probe molecule, produces the *extended surface* of the model. This display is useful for studying intermolecular contacts. If the user brings two models together—one with extended surface displayed, the other as a simple stick model—the points of intermolecular contact are where the extended surface of one model touches the atom centers of the second model.

The color of the displayed surface is usually the same as the color selected for the underlying atoms. In Plate 18, for example, oxygens of one of the heme carboxyl groups produces the large red bulge of accessible surface (the other carboxyl is hydrogen bonded to serine-64 and is much less accessible to solvent). Alternatively, color can reflect surface charge (commonly, blue for positive, red for negative, with lighter colors for partial charges) or surface polarity (contrasting colors for hydrophobic and hydrophilic regions). These displays facilitate finding regions of the model to which ligands of specified chemical properties are likely to bind.

I. Exploring intermolecular interactions: Multiple models

Formulating proposed mechanisms of protein action requires investigating how proteins interact with ligands of all kinds, including other proteins. Molecular modeling programs allow the user to display and manipulate several models, either individually or together. With SPV and many other modeling programs, the number of models is limited only by computer memory and speed. Tools for this purpose usually allow all of the same operations as the viewing tools but permit selection of models affected by the operations. In *docking experiments* (a term taken from satellite docking in the space program), one model can be held still while another is moved into possible positions for intermolecular interaction. Labeling, measurement, and surface tools are used simultaneously during docking to ensure that the proposed interactions are chemically realistic. Some programs include computational docking, in which the computer searches for optimal interaction, usually from a user-specified starting point.

J. Displaying crystal packing

Many molecular modeling programs include the capacity to display models of the entire unit cell. All the program needs as input is a set of coordinates for one molecule, the unit-cell dimensions, and a list of equivalent positions for the crystal space group. The user can display one cell or clusters (say, $2 \times 2 \times 2$) of cells. The resulting images, particularly when teamed with surface displays, reveal crystal-packing interactions, allowing the user to see which parts of the crystallographic model might be altered by packing, and might thus be different from the solution structure. For examples of crystal-packing displays, see Chapter 4, Section II.H, Figs. 4.15 and 4.16. Unit-cell tools usually allow the user to turn equivalent positions on and off individually, making them useful for teaching the topics of equivalent positions and symmetry. For example, SPV allows the user to create new models by specifying symmetry operations or selecting them from a list or from symmetry lines in PDB files.

K. Building models from scratch

Iin addition to taking coordinate files as inputs, modeling programs allow the user to build peptides to specification and to change amino-acid residues within a model. To build new models, users select amino acids from a palate or menu and direct the program to link the residues into chains. Users can specify conformation for the backbone by entering backbone angles Φ and Ψ, by selecting a common secondary structure or by using the tools described earlier for exploring structural change. Model-builder tools are excellent for making illustrations of common structural elements like helices, sheet, and turns. SPV allows you to start building a model by simply providing a text file of the desired sequence in one-letter abbreviations; it imports the sequence and models it as an alpha helix for compactness. Then you can select parts of the model and either select their secondary structural type or set their Ramachandran angles. Alternatively, you can display a Ramachandran diagram for the model and change torsional angles by dragging residue symbols to new locations on the diagram, watching the model change as you go. Similar tools are used to replace one or more side chains in a model with side chains of different amino acids, and thus explore the local structural effects of mutation.

IV. Other tools for studying structure

It is beyond the scope of this little book to cover all the tools available for studying protein structure. I will conclude by listing and briefly describing additional tools, especially ones used in conjunction with modeling on graphics systems.

A. Tools for structure analysis

In addition to molecular graphics, a complete package of tools for studying protein structure includes many accessorry programs for routine structure analysis. The chores executed by such programs include the following:

- Calculating Φ and Ψ angles and using the results to identify elements of secondary structure as well as to display a Ramachandran diagram, which is useful in finding model errors during structure refinement (crystallography or NMR) or homology modeling. As I mentioned earlier, SPV has a unique interactive Ramachandran plot window that allows the user to change main-chain conformational angles in the model (Plate 16).
- Using distance and angle criteria to search for hydrogen bonds, salt links, and hydrophobic contacts, and producing a list of such interactions. SPV calculates and displays hydrogen bonds according to user specifications of distance and angle. A very powerful pair of menu commands in SPV allows you to show hydrogen bonds only from selected residues or hetero groups and then to show only residues with visible hydrogen bonds. In these two operations, you can eliminate everything from the view except, for example, a prosthetic group and its hydrogen-bonding neighbors.
- Comparing homologous structures by least-squares superposition of one protein backbone on another. The result is a new coordinate set for one

model that best superimposes it on the other model. I used such a tool in SPV to compare the X-ray and NMR structures of thioredoxin in Plate 5. SPV provides superposition tools combined with sequence comparison to improve the structural alignment between proteins that are only moderately homologous. In addition, you can color a protein by the rms deviation of its atoms from those of the reference protein on which it was superimposed, giving a vivid picture of areas where the structures are alike and different. For example, in Plate 19, full backbone models of the X-ray and NMR structures (PDB 1ert and 3trx) are superimposed. The X-ray model is gray, and the NMR model is colored by rms deviations of corresponding residues from the X-ray model. Residues for which the two models deviate least are colored blue. Those exhibiting the greatest deviations are red. Residues with intermediate deviations are assigned spectral colors between blue and red. Even if the X-ray model were not shown, it would be easy to see that the most serious disagreement between the two models lies in the surface loop at the lower right, and that the two models agree best in the interior residues.

- Calculating surface electric fields, which can be displayed in graphics programs to reveal regions that would attract molecules of opposites charge or to show expected direction of movement of charged ligands.
- Building additional subunits using symmetry operations, either to complete a functional unit (Chapter 8, Section II.D) or to examine crystal packing. SPV allows you to build additional subunits by selecting symmetry operations from a palette, clicking on symmetry operations listed in a PDB file, or typing in the components of a transformation matrix.
- Carrying out homology modeling. As mentioned in Chapter 10, SPV is a full homology modeling program. Using SPV, a web browser, and electronic mail, you can obtain template files from SWISS-PROT, align, average, and thread your sequence onto the template, build loops or select them from a database, find and fix clashes, submit modeling projects to SWISS-MODEL, and retrieve them to examine the results or apply other modeling tools. GROMOS energy minimization is currently being added to SPV. Plate 20 shows a homology model started in SPV and completed at SWISS-MODEL. The homology model is shown as a ribbon colored according to model B-factors (Chapter 10, Section III.D). Two templates used in the modeling are shown as black and gray alpha-carbon displays. (For more information about these proteins, see the legend to Plate 20.)
- Displaying an electron-density map and adjusting the models to improve its fit to the map (see Plate 21 and the cover of this book). SPV can display maps of several types (CCP4, X-PLOR, DN6). I am aware of no programs currently available for computation of maps from structure factors on personal computers, but I am sure this will soon change.

Structure factors are available for some of the models in the PDB, and can usually be obtained from the depositors of a model.

B. Tools for modeling protein action

The crystallographic model is used as a starting point for further improvement of the model by energy minimization and for simulations of molecular motion. Additional insight into molecular function can be obtained by calculating charge densities and bond properties by molecular orbital theory. For small molecules, some of these calculations can be done "on the fly" as part of modeling. For the more complex computations, and for larger molecules, such calculations are done outside the graphics program, often as separate tasks on computers whose forte is number crunching rather than graphics. SPV can write files for export to several widely used programs for energy minimization and molecular dynamics. But as personal computers get faster, these operations will no longer require transfer to specialized machines.

V. A final note

Making computer images and printed pictures of molecular models endows them with the concreteness of everyday objects. While exploring models, viewers can easily forget the difficult and indirect manner by which they are obtained. I wrote this book in hopes of providing an intellectually satisfying understanding of the origin of molecular models, especially those obtained from single-crystal X-ray crystallography. I also hope to encourage readers to explore the many models now available, but to approach them with full awareness of what is known and what is unknown about the molecules under study. Just as good literature depicts characters and situations in a manner that is "true to life," a sound model depicts a molecule in a manner that is true to the data from which it was derived. But just as real life is more multifarious than the events, settings, and characters of literature, not all aspects of molecular truth (or even of crystallographic, NMR, or modeling truth) are reflected in the colorful model floating before us on the computer screen. The user must probe more deeply into the esoterica of structure determination to know just where the graphics depiction is not faithful to the data. The user must probe further still—into the wider literature on the molecule—to know whether the model is faithful to other evidence about structure and action. The conversation between structural models and evidence on all sides continually improves models as depictions of molecules.

If I have stimulated your interest in crystallography itself, you may be wondering where to go from here. As your next step toward a truly rigorous understanding of crystallography, I suggest *Crystal Structure Analysis for Chemists and Biologists: Methods in Stereochemical Analysis* by Jenny Glusker, Mitchell Lewis, and Miriam Rossi (John Wiley & Sons, 1994); *X-ray Structure Determination: A Practical Guide*, 2nd edition, by George H. Stout and Lyle H. Jensen (John Wiley and Sons, Inc., 1989); and *Practical Protein Crystallography*, by Duncan E. McRee (Academic Press, Inc., 1993). On the World Wide Web, I recommend *Crystallography 101* by Bernhard Rupp. You will find links to this and other crystallography resources at the CMCC Home Page, www.usm.maine.edu/~rhodes/CMCC.

Index

Viewing Stereo Images

To see three-dimensional images of these models, use a stereo viewer such as item #D8-GEO8570, Carolina Biological Supply Company (1-800-334-5551). You can view stereo pairs without a viewer by training yourself to look at the left image with your left eye and the right image with your right eye. This is neither as difficult nor as strange at it sounds. (According to my ophthalmologist, it is not harmful to the eyes, and may in fact be good exercise for eye muscles.) Try putting your nose on the page between the two views. With both eyes open, you will see the two images superimposed, but out of focus, because they are too close to your eyes. Slowly move the paper away from your face, trying to keep the images superimposed until you can focus on them. (Keep the line between image centers parallel to the line between your eyes.) When you can focus, you will see three images. The middle one should exhibit convincing depth. Try to ignore the flat images on either side. This process becomes easier with practice. You may find it helpful to try this process first on an image of solid objects, such as Plate 4, or with a very simple image, such as Plate 6. For more help with viewing stereo images in books or on computers, go to the CMCC Home Page, www.usm.maine.edu/~rhodes/CMCC.

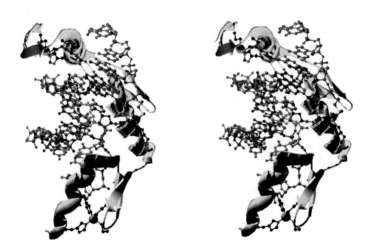

Plate 1 Stereo photograph of Zif268/DNA complex (see N. P. Pavletich and and C. O. Pabo, *Science* **252**, 809, 1991). (For discussion, see Chapter 1.) NOTE: Atomic coordinates for preparing this display were obtained from the Protein Data Bank (PDB), which is described in Chapter 7. The PDB file code is 1zaa. To expedite your access to all models shown in this book, I provide file codes in this format: PDB 1zaa. Image created by Swiss-PdbViewer, rendered by POV-Ray. To obtain these programs, see the CMCC Reader's Page.

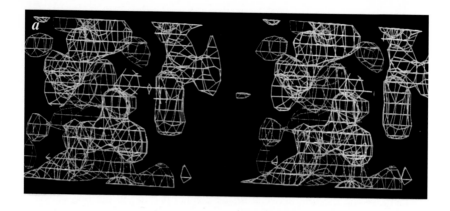

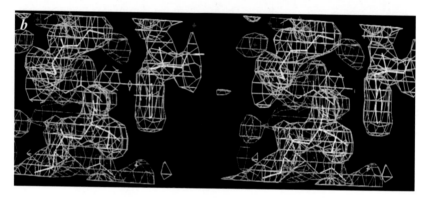

Plate 2 (*a*) Small section of molecular image displayed on computer graphics terminal. (*b*) Image (*a*) is *interpreted* by building a molecular model within the image. Computer graphics programs allow parts of the model to be added and their conformations adjusted to fit the image. The protein shown here is adipocyte lipid binding protein (ALBP, PDB 1alb). (For discussion, see Chapter 2.)

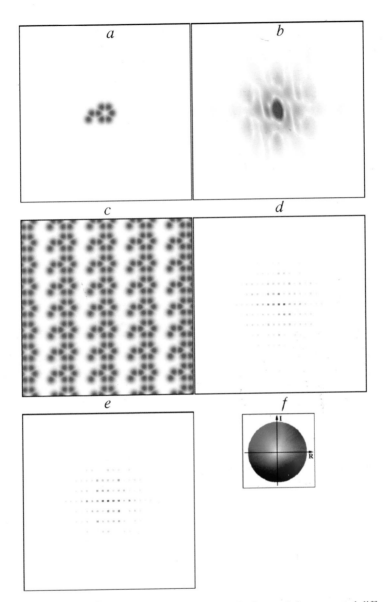

Plate 3 Simple asymmetric object, alone and in a lattice, and the computed diffraction pattern of each, with phases depicted by color. Darkness of color indicates the intensity of a reflection. The phase angle of a region in *b* or a reflection in *d* corresponds to the angle of its color on the color wheel *f*. Experimental diffraction patterns do not contain phase information, as in *e*. (For discussion, see Chapter 2.) Images computed and generously provided by Dr. Kevin Cowtan. For more illustrations of Fourier transforms as they apply to macromolecular crystallography, direct your web browser to the CMCC Home Page (www.usm.maine.edu/ ~rhodes/CMCC) and select Kevin Cowtan's *Book of Fourier*.

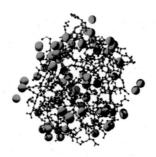
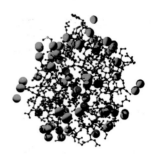

Plate 4 One molecule of crystalline adipocyte lipid-binding protein (ALBP, PDB 1alb), showing ordered water molecules on the surface and within a molecular cavity where lipids are usually bound. Protein is shown as a ball-and-stick model with carbon dark gray, oxygen red, and nitrogen blue. Ordered water molecules, displayed as space-filling oxygen atoms, are green. (For discussion, see Chapter 3.). Image: SPV/POV-Ray.

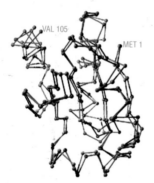
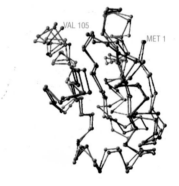

Plate 5 Models of the protein thioredoxin (human, reduced form) as obtained from x-ray crystallography (blue, PDB 1ert) and NMR (red, PDB 3trx). Only backbone alpha carbons are shown. The models were superimposed by least-squares minimization of the distances between corresponding atoms, using Swiss-PdbViewer. (For discussion, see Chapter 3.) Image: SPV/POV-Ray.

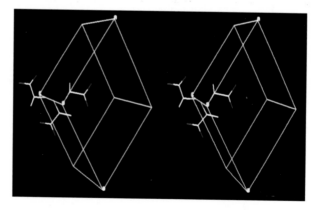

Plate 6 Threefold screw axis (3_1). (For discussion, see Chapter 4.)

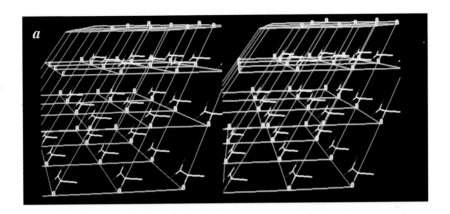

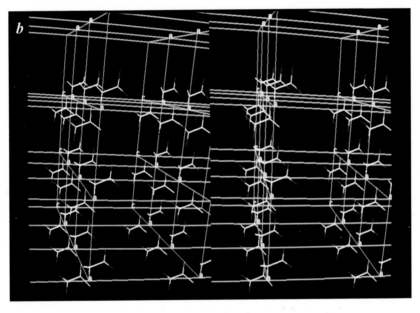

Plate 7 Alanine in hypothetical (*a*) *P*1 and (*b*) *P*2$_1$ unit cells. (For discussion, see Chapter 4.)

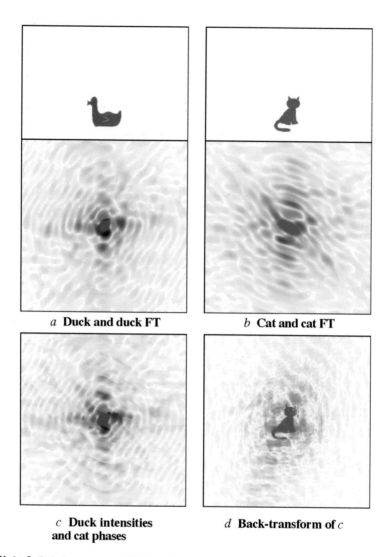

a **Duck and duck FT** *b* **Cat and cat FT**

c **Duck intensities and cat phases** *d* **Back-transform of** *c*

Plate 8 Relative amounts of information contained in reflection intensities and phases. The Fourier transforms of duck and cat are shown in (*a*) and (*b*). In (*c*), the intensity or shading of the duck transform is combined with the phases or colors of the cat transform. The back-transform of (*c*), shown in (*d*), contains a clear image of the cat. Ironically, the phases, which are so difficult to obtain, provide more information than the easily obtained intensities. (For discussion, see Chapter 6.) Figure generously provided by Dr. Kevin Cowtan.

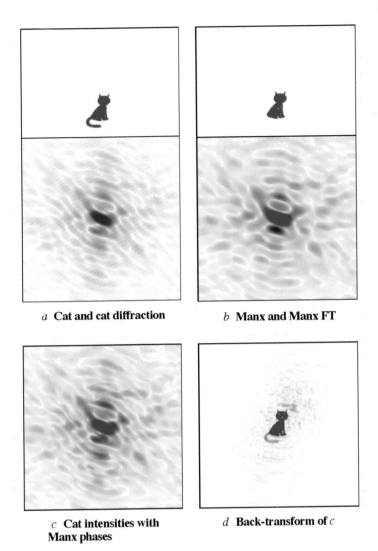

a **Cat and cat diffraction**　　　*b* **Manx and Manx FT**

c **Cat intensities with**　　　*d* **Back-transform of** *c*
Manx phases

Plate 9 Structure determination by molecular replacement. We know the structure of the manx cat, and want to learn the structure of the cat. In (*a*) the cat (unknown structure) is shown, along with its Fourier transform without phases. This transform is analogous to a set of measured diffraction intensities. In (*b*), the manx cat (known structure and source of starting phases) is shown with its Fourier transform, including phases. Calculating transforms from known models gives both intensities and phases. In (*c*), the phases from the known model (*b*) are added to the intensities of the unknown (*a*). The back-transform of (*c*), shown in (*d*), shows a weak but distinct image of the cat's tail, the only structural difference between the known and unknown structures. This shows that intensities contain enough information to reveal differences between similar structures, and to allow a similar structure to be used as a phasing model. (For discussion, see Chapter 6.) Figure generously provided by Dr. Kevin Cowtan.

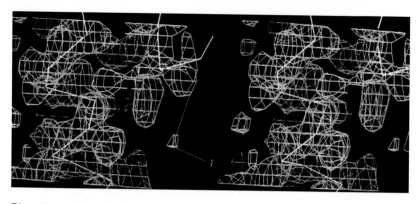

Plate 10 Alpha-carbon model of ALBP built into electron-density map. (For discussion, see Chapter 7.)

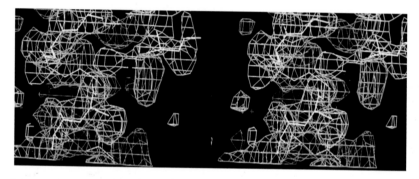

Plate 11 Polyalanine model of ALBP built into electron-density map. This section of the final ALBP model is shown in Plate 2. (For discussion, see Chapter 7.)

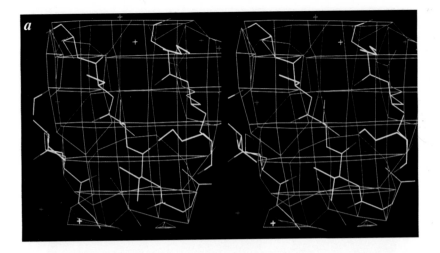

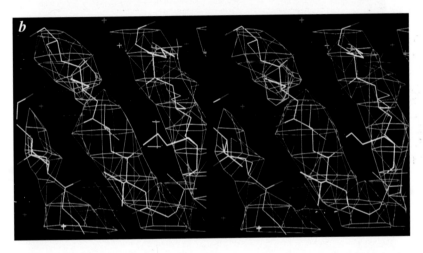

Plate 12 Electron-density maps at increasing resolution. Maps were calculated using final phases, and Fourier series were truncated at the resolution limits indicated: (*a*) 6.0 Å; (*b*) 4.5 Å; (*c*) 3.0 Å; (*d*) 1.6 A. (For discussion, see Chapter 7.) (*Continues*)

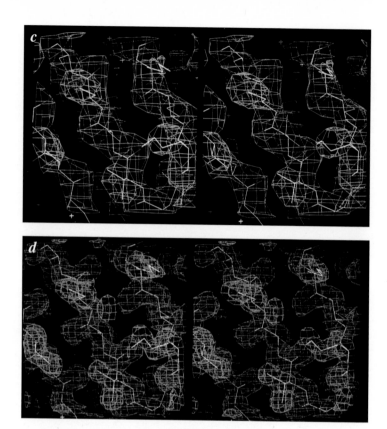

Plate 12–*Continued*

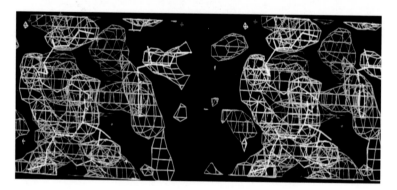

Plate 13 ALBP electron-density map calculated with molecular-replacement phases before any refinement, shown with the final model. Compare with Plate 2, which shows the final electron-density map in the same region. (For discussion, see Chapter 8.)

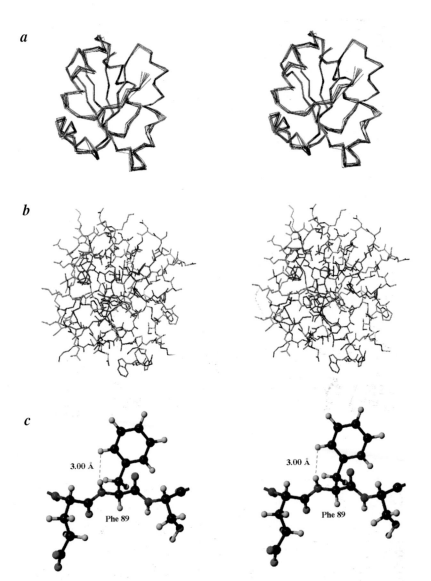

Plate 14 (*a*) Ten of the 33 NMR models of thioredoxin (PDB 4trx). (*b*) Averaged model (PDB 3trx), colored by rms deviation of atom positions in the ensemble from the average position. For each residue, main-chain colors reflect the average rms deviation for C, O, N, and CA, and all side-chain atoms are colored to show the average rms deviation for atoms in the whole side chain. (*c*) Detail of the averaged model at phenylalanine-89, showing the averaged distance between the two hydrogens involved in the distance restraint indicated as "F89N,δ" in the NOESY spectrum, Figure 10.6. (For discussion, see Chapter 10.) Image: SPV/POV-Ray.

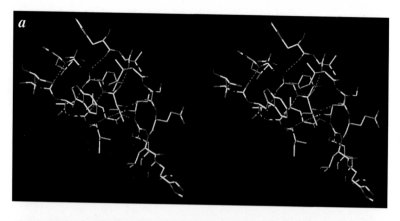

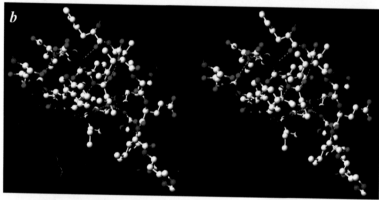

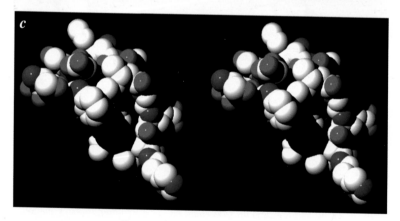

Plate 15 A selection of common types of computer graphics models, all showing the same three strands of pleated-sheet structure from cytochrome *b*5 (PDB 3b5c). (*a*) Wireframe; (*b*) ball and stick; (*c*) space filling; (*d*) ribbon backbone with ball-and-stick side chains. (For discussion, see Chapter 11.) Image: SPV/POV-Ray. (*Continues*)

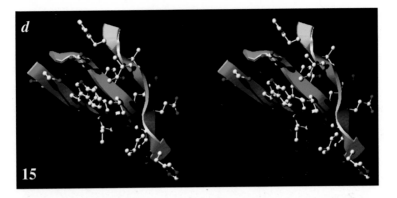

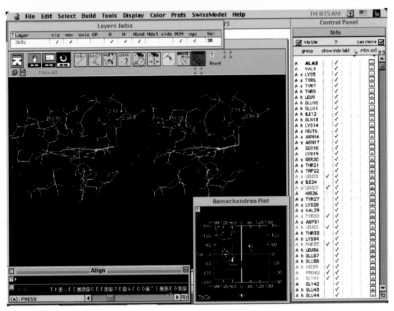

Plate 16 Screen shot of Swiss-PdbViewer in use on a Power Macintosh computer. Controls for manipulating the model are at the top of the main graphics window, which can be expanded to fill the screen. The Control Panel lists residues in the model and allows selection of residues for display, coloring, labeling, and surface displays. The Align window shows residue sequence also, including alignment of multiple models if present. The Ramachandran Plot window shows main-chain torsional angles for residues currently selected (purple in Align window and red in Control Panel). Dragging dots on the Rama plot changes torsion angles interactively. The Layers Infos window allows control of display features for multiple models in any combination. In the graphics window is a stereo display of the heme region of cytochrome $b5$ (PDB 3b5c). Selected hydrogen bonds are shown in green, and measured distances are shown in yellow. A torsion operation is in progress (dark button, top of graphic window). The side-chain conformation of phenylalanine-58 is being changed. Clashes between the side chain and other atoms are shown in pink. The user can read the PDB file of the currently active model by clicking the document icon at the upper right of the graphics window. Clicking a residue in the PDB file centers the graphics model on that residue. (For discussion, see Chapter 11.)

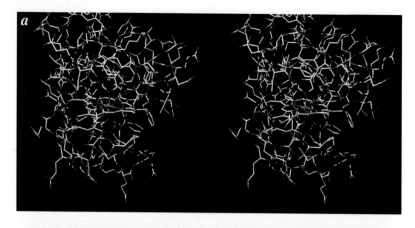

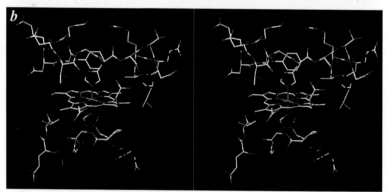

Plate 17 Heme region of cytochrome *b*5 (PDB 3b5c). (*a*) View without clipping; (*b*) same view after "slab" command to eliminate all except contents of a 12-Å slab in the *z*-direction. (For discussion, see Chapter 11.) Image: SPV/POV-Ray.

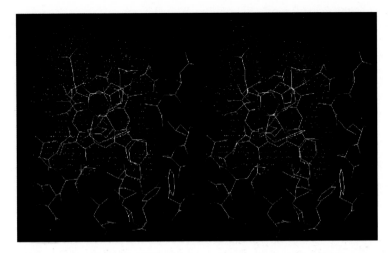

Plate 18 Dotted surface displays of heme in cytochrome *b*5 (PDB 3b5c). Smaller van der Waals surface encloses heme completely. Small outer dotted surface is the solvent-accessible surface of the heme. Most of the heme surface is buried within the protein. (For discussion, see Chapter 11.) Image: SPV/POV-Ray.

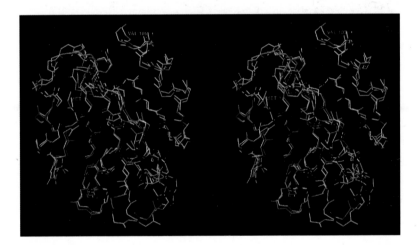

Plate 19 X-ray and NMR models of thioredoxin (PDB 1ert and 3trx) superimposed by least-squares fitting of corresponding alpha carbons in the two models. The x-ray model is gray. Residues of the NMR model are colored according to rms differences in atom positions between the two models. Residues with smallest deviations are blue, those with largest deviations are red, and those of intermediate deviations are in spectral colors between blue and red. (For discussion, see Chapter 11.) Image: SPV/POV-Ray.

Plate 20 Homology modeling project returned from SWISS-MODEL. The target protein is a fragment of FasL, a ligand for the widely expressed mammalian protein Fas. Interaction of Fas with FasL leads to rapid cell death via apoptosis. The template proteins are (1) tumor necrosis factor receptor P55, extracellular domain (PDB 1tnr, black) and (2) tumor necrosis factor-alpha (PDB 2tun, gray). The modeled FasL fragment is shown as ribbon and colored by model β-factors. Only the alpha carbons of the templates are shown. (For discussion, see Chapter 11.) Image: SPV/POV-Ray.

Plate 21 Model and portion of electron-density map of bovine Rieske iron-sulfur protein (PDB 1rie). The map is contoured around selected residues only. (For discussion, see Chapter 11.) Image: SPV/POV-Ray.